The Wheel

COLUMBIA STUDIES IN INTERNATIONAL
AND GLOBAL HISTORY

COLUMBIA STUDIES IN INTERNATIONAL AND GLOBAL HISTORY

The idea of "globalization" has become a commonplace, but we lack good histories that can explain the transnational and global processes that have shaped the contemporary world. Columbia Studies in International and Global History encourages serious scholarship on international and global history with an eye to explaining the origins of the contemporary era. Grounded in empirical research, the titles in the series transcend the usual area boundaries and address questions of how history can help us understand contemporary problems, including poverty, inequality, power, political violence, and accountability beyond the nation-state.

Cemil Aydin, *The Politics of Anti-Westernism in Asia: Visions of World Order in Pan-Islamic and Pan-Asian Thought*

Adam M. McKeown, *Melancholy Order: Asian Migration and the Globalization of Borders*

Patrick Manning, *The African Diaspora: A History Through Culture*

James Rodger Fleming, *Fixing the Sky: The Checkered History of Weather and Climate Control*

Steven Bryan, *The Gold Standard at the Turn of the Twentieth Century: Rising Powers, Global Money, and the Age of Empire*

Heonik Kwon, *The Other Cold War*

Samuel Moyn and Andrew Sartori, eds., *Global Intellectual History*

Alison Bashford, *Global Population: History, Geopolitics, and Life on Earth*

Adam Clulow, *The Shogun and the Company: The Dutch Encounter with Tokugawa Japan*

The Wheel

Inventions & Reinventions

Richard W. Bulliet

Columbia University Press
New York

Columbia University Press and the author wish to express their appreciation for assistance given by an anonymous donor toward the cost of publishing this book.

Columbia University Press
Publishers Since 1893
New York   Chichester, West Sussex
cup.columbia.edu

Library of Congress Cataloging-in-Publication Data
Bulliet, Richard W., author.
    The wheel : inventions and reinventions / Richard W. Bulliet.
        pages cm. — (Columbia studies in international and global history)
    Includes bibliographical references and index.
    ISBN 978-0-231-17338-4 (cloth : alk. paper)
    ISBN 978-0-231-54061-2 (e-book)
    1. Wheels—History.   I. Title.
    TJ181.5.B85 2015
    621.8—dc23

                                                    2015020787

Columbia University Press books are printed on permanent
and durable acid-free paper.
This book is printed on paper with recycled content.
Printed in the United States of America

c 10 9 8 7 6 5 4 3 2 1

COVER IMAGE: (*top*) Bridal carriage (Collection of the Kunstammlungen der Veste Coburg, Coburg, Germany; (*middle*) model of Chinese chariot (DEA / E. LESSING / Granger, NYC—All rights reserved); (*bottom*) Benz Patent Motorwagen (Flickr.com)
COVER DESIGN: Jordan Wannemacher

References to websites (URLs) were accurate at the time of writing. Neither the author nor Columbia University Press is responsible for URLs that may have expired or changed since the manuscript was prepared.

published with the aid of a grant from
Figure Foundation

coded by idea rolls the self autograph

To my son and editor

Mark Bulliet

Contents

Acknowledgments

Given, as I am, to doing my thinking out loud, I am particularly grateful to the three undergraduates who took my seminar on wheeled transportation in 2012: John Clay Evans, Kevin K. O'Connell, and Angelina Sapienza. As my ideas began to jell, I had three opportunities to experiment with them in lecture format. My thanks go to Professor James A. Riley, who was instrumental in my being invited to present the F. E. L. Priestley Lectures in the History of Ideas at the University of Toronto in 2012; to my old friend Professor Roy Mottahedeh, who facilitated my giving the Richard N. Frye Lecture at Harvard in 2013; and to Professor Wallace Broecker, who, at the suggestion of Professor David Ho of the University of Hawai'i, invited me to speak at the Earth Science Colloquium of the Lamont-Doherty Earth Observatory of Columbia University in 2013. A number of good and patient friends listened to my monologues and read the manuscript as it began to emerge. Notable among them are Jahan Salehi, Ramzi Rouighi, Howard Shawn, and Hossein Kamaly. As the text finally came together, my son Mark repeatedly urged me to continue rewriting and did two incisive edits of the entire manuscript, from which I benefited more than I can say. Finally, Irene Pavitt markedly improved the manuscript with her expert editing, and Karen Ehrmann did a superb job of researching images and obtaining permissions.

The Wheel

Wheel Versus Wheel

n 1850, the steam engine was ranked as the world's greatest invention. By 1950, the wheel, a much older invention, had surpassed it. The advent of the electric motor and internal-combustion engine partly explains the decline of the steam engine; but the spread of automobiles, trucks, and buses—not to mention grocery carts, bicycles, and roll-aboard luggage—played a greater role. For in 1850, the wheeled vehicles that rumbled over the cobblestones of city streets and jounced along the rutted dirt roads of the countryside seemed neither new nor particularly ingenious.

The question of when and where the wheel first appeared did not excite much interest before the twentieth century, and even now it is hard to find comprehensive accounts of how wheels have been used for transportation in different times and places. Nor has it been recognized that wheels come in three strikingly different forms, each with its own history of invention. Two kinds, wheels that are fixed to the ends of an axle and turn in unison as the axle turns—the whole apparatus is called a wheelset (figure 1)—and wheels that rotate independently from each other on the ends of a stationary axle, date to between 4000 and

FIGURE 1 Railroad wheelsets. (Colourbox.com)

3000 B.C.E. They have thus been around for more than five thousand years. The third form, termed a caster, rotates on an axle and also pivots in a socket situated above it, the axis of pivoting being somewhat offset from the axis of rotation. Despite its technological simplicity, the caster seems to have come into use only three hundred years ago, or five thousand years after the other two forms.

We encounter wheels of all three designs almost daily. Railroad and subway cars originally rolled on wheelsets, whereas road vehicles use wheels that rotate individually, mostly developed from an original form of two wheels rotating independently on the ends of an axle. And casters abound on wheelchairs, gurneys, laundry carts, dollies, baby strollers, and desk chairs.

Steering is key to the three different wheel designs. Locomotives and railcars cannot make sharp turns because in the process of turning, the outer wheel covers a greater distance than the inner wheel and thus must make more rotations. Consequently, when the wheels turn in unison in a wheelset, the outer wheel is skidded or dragged around the turn. Railroad steering is accomplished by following rails, and railroads are designed with only very gradual curves.

Axles with independently rotating wheels do not have this problem, but a wagon equipped with four wheels on two axles still has a hard time turning unless the front wheels can change direction. Pivoting a front axle in its middle underneath a wagon was the most common way of changing the direction of the wheels until two hundred years ago, when alternatives were devised to allow each front wheel to pivot separately. Automobiles use these later modifications and thus can turn much sharper corners than can railroad cars.

As for the caster, if pressure comes from the side or from the person pushing the vehicle applying more force on one side of the handle (or on one handle if there are two) than on the other, the wheels turn automatically in the desired direction. Grocery carts and wheelchairs typically combine casters in front with rear wheels that do not pivot, but dollies and desk chairs use only casters and can easily move in any direction.

Since the notion that the wheel may be humankind's greatest invention has arisen quite recently, I shall begin this exploration of the invention of the wheel with the modern era of motorized transport. Underlying this history is a competition between wheelsets in the form of railroads and independently rotating wheels in the form of automobiles. Once the importance to the world of today of this wheel-versus-wheel competition has been established, I will go back in time to explore the invention of the first wheels and examine the crucial episodes in wheel history that culminated in the modern transition from animal power to motorized transport, on the one hand, and to human power, as represented by the history of the rickshaw and the caster, on the other.

At the dawn of the European Renaissance in the fourteenth century, four-wheeled vehicles were not in common use anywhere in the world. Two-wheeled carts were easier to build, easier to steer, and mechanically more efficient, since two wheels turn with only half the friction of four. Even in Europe, the only world region that never abandoned the four-wheel concept, passenger wagons were used almost exclusively by noblewomen, the infirm, and the elderly. Noblemen rode horses as a sign of their elite warrior status, while eminent churchmen rode mules.

This situation changed dramatically over the next three centuries in what we will call the carriage revolution. By 1650, from the Thames to the Danube, the word "coach," in various spellings, had become the general European term for a four-wheeled passenger vehicle, synonymous in English with the word "carriage." The number of coaches soared. The number of carriages in Britain, for example, could be counted on one hand in 1570, while the official tax totals in 1814 enumerated "23,400 four-wheeled vehicles paying duty to the government; 27,300 two-wheel, and 18,500 tax carts [inexpensive vehicles taxed at a lower rate]; a total of 69,200," which amounted to one vehicle for every 145 inhabitants.[1]

As the numbers grew, so did the variety. European carriage makers and wheelwrights catered to every taste. In addition to generic conveyances like "long wagons" for hauling freight and stagecoaches for transporting passengers, they turned out a dazzling array of barouches, berlins, caroches, *carosses*, chaises, clarences, coupes, diligences, *ducs*, fiacres, gazelles, hackneys, jaggers, milords, phaetons, rockaways, *reise-pritschkas*, socials, spiders, surreys, and victorias—just to mention designs of the four-wheeled variety. The differences among them involved size, weight, type of springs or suspension, steering mechanism, driver position, passenger capacity, seating arrangement, number of horses, and a galaxy of door, window, and top designs. By the mid-nineteenth century, the range of choices available had turned the carriage, or buggy (to use the American term), into a perfect billboard for displaying one's economic status or personal style, thus prefiguring the automobile culture that was to evolve in the twentieth century and that continues into the twenty-first.

In addition, the Europeans and Americans who made up the social elite in overseas colonies, or whose diplomatic or business duties compelled them to sojourn in non-Western lands, often used carriages to advertise the cultural and technological superiority of their home countries. Imitating a fellow Englishman who gained fame writing about his globetrotting excursions as the "Blind Traveller," John Kitto billed himself as the "Deaf Traveller" and remarked that in 1833 he "saw no wheel-carriages of any kind in Persia."[2] Exactly fifty years later, however, the first American minister resident in Iran, Samuel G. W. Benjamin, reported that "the number of carriages owned by Persian and European gentlemen [in Tehran] is nearly 500, all imported."[3]

Coach building far outstripped road building, however. The problem of finding a road surface that would bear up under heavy traffic and iron-rimmed wheels was paralleled by the problem of finding a way to pay for road maintenance. Until new experiments with road building began to bear fruit in the mid-nineteenth century, the surfaces beneath the carriage wheels remained rutted, muddy, and poorly paved—if paved at all. This was particularly true in the countryside, but miserable roads existed even in major cities. In 1703, for example, during a trip south from London to Petworth, fifty miles away, the carriage carrying the Habsburg emperor Charles VI overturned twelve times on the road.[4] And a half century later, Mile End Road, the major thoroughfare leading east from the entrance to the City of London at Aldgate, was described as "a stagnant lake of deep mud from Whitechapel to Stratford," a distance of four miles.[5]

Roads outside Europe were at least as bad. In Washington, D.C., for example, the track between the Capitol and Thomas Jefferson's White House, a way labeled Pennsylvania Avenue on the map, was so crude that "every turn of your wagon wheel . . . is attended with danger."[6] And the stagecoach that connected Capitol Hill with Georgetown, just five miles away, made a round trip twice a day "in a halo of dust . . . pitch[ing] like a ship in a seaway, among the holes and ruts of this national highway."[7]

To this may be added the steadily growing problem of horse manure blighting pedestrian activity. Ironically, as the condition of streets and roads became better, the accumulation of manure became worse. One estimate from the 1890s predicted that by 1950 London would be buried in nine feet of horse droppings, while another concluded that by 1930 manure would reach the third-story windows of buildings in Manhattan.

To begin to understand the evolution of modern motor transport from the vantage point of the wheel, we must consider the alternatives offered by the emergence of two different wheel technologies in the fourth millennium B.C.E., for the choice between wheels fixed to axles and independently rotating wheels that originated in that era reemerged in the industrial age to affect modern transport at a fundamental level. In the chapters that follow, I propose that the wheelset, or fixed-wheel design, was invented first, perhaps as early as 4000 B.C.E., for use in copper mines situated in eastern Europe's Carpathian Mountains, which stretch westward from Romania, across western Ukraine, southern Poland, and northern Hungary, to Slovakia. The independently rotating–wheel design, very likely inspired by the wheelset, came on the scene a few centuries later, somewhere in southeastern Europe between the Carpathians and the broad plain north of the Black Sea, where people speaking Proto-Indo-European shared a set of technical terms for wagons with rotating wheels, but did not have a word for the wheelset.

In the absence of any archaeological traces of roads, we can assume that most wagons with fully rotating wheels moved cross-country over uneven dirt and grass. This would have required wheels of fairly large diameter, around thirty inches to judge from surviving examples. The stone floors of mines, however, could be as smooth and as straight as the miners chose to make them. Yet ore is extremely heavy, so the four-wheeled mine-cars pushed by the miners were no bigger than today's wheeled laundry carts and had comparatively small wheels, perhaps around one foot across. Given straight tunnels, flat stone floors, and the absence of draft animals, there was little need for steering. And it was better to use a wheelset than an axle with independently rotating wheels because when the former broke, the mine-car simply stopped; but having a single wheel come off, as commonly happened with rotating wheels, could dump the entire load of ore onto the tunnel floor.

Mine-cars continued to be pushed by European miners down to the eighteenth century, when pit ponies and mules took over some of the work in Britain and elsewhere. Also by then, the idea had spread of using wooden tracks—later replaced by iron—to guide the cars and provide a smoother surface for them to roll on. This innovation arose in the sixteenth century, if not earlier.

Outside the mining world, fixed-wheel vehicles were rare, though two-wheeled versions saw some use in the Roman Empire, and rustic oxcarts of this design could occasionally be found in southern Europe from Portugal to Turkey in the early twentieth century and were not unknown in India. Although the outer wheel had to be skidded around sharp corners, carts with solid fixed wheels were easier to build and more durable than those with independently rotating wheels, especially if the rotating wheels had spokes. Nevertheless, the two-wheeled, animal-drawn cart with independently rotating wheels was by far the most common form of wheeled transport, not just in Europe but from India and China to South America.

Soon after Thomas Newcomen's steam engine burst on the scene around 1710, innovative thinkers began to imagine shrinking the massive, four-story-tall device that Newcomen had designed for pumping water out of flooded mines down to a size that could be put on wheels. The vehicular designs they had to work with were three: the two-wheeled cart and the four-wheeled wagon or carriage, both of which were equipped with independently rotating wheels; and the four-wheeled mine-car, which rode on wheelsets and rails.

Since steam engines, no matter how compactly designed, seemed intrinsically to be constructions of great weight, balancing one on two wheels and a single axle was out of the question. In fact, no motor-vehicle design based on two wheels placed side-by-side saw the light of day until Dean Kamen unveiled the Segway in 2001.[8] As an alternative, however, it seemed feasible to add to a conventional cart design a pivoting third wheel placed either before or behind the two main wheels. This might not only help distribute the weight of the steam engine, but also facilitate steering. Nicholas-Joseph Cugnot pursued this idea with a vehicle that he built for the French army in 1769 (figure 2). To achieve a weight balance between the steam engine and the artillery or other heavy load that the vehicle was intended to pull, Cugnot put the boiler far forward and the steam engine itself directly over a single front wheel

FIGURE 2 A reconstruction of Nicholas-Joseph Cugnot's
three-wheeled steam-powered wagon, originally built in 1769.
(Photograph by Andrew Duthie. Musée des Arts et Métiers, Paris)

that could be turned, if one had sufficient strength, by means of a two-handed steering tiller. When the army rejected the design, the idea of balancing a steam engine on a single wheel died with it.

In 1803, however, Richard Trevithick, one of the pioneers of steam-engine and locomotive design, tried a different idea. His London Steam Carriage used the third wheel solely for steering. The weight of the engine and of the stagecoach-like passenger compartment rested on two huge rear wheels (figure 3). The driver sat in front and steered the pivoting front wheel with a long, angled tiller of the type that had been in use for fifty years on three-wheeled Bath chairs at England's seaside resorts. (The Bath chair's tiller not only steered a small front wheel, but was hinged so that it could either extend forward as a handle for an attendant to pull or flip back to give steering control to the rider while the attendant pushed the chair from behind [figure 4].) Although Trevithick's contraption proved too uncomfortable and expensive for practical use,

FIGURE 3 Richard Trevithick's London Steam Carriage, demonstrated in London in 1803. (Wikipedia Commons)

FIGURE 4 The Bath chair, which was steered by means of a tiller. (The Florida Center for Instructional Technology, University of South Florida, Tampa)

FIGURE 5 H. T. Alken, *A View in Whitechapel Road* (1831), a caricature of the
steam-powered cars of the future. (Wikipedia Commons)

his vision aroused enough interest to spur a satirist to imagine what
steam traffic on a London street might look like in the future (figure 5).

The motorized-tricycle concept did not die with the London Steam
Carriage. In 1867, a practical steamroller went into production. Its engine's great weight, which was an asset rather than a liability given the
vehicle's purpose, was divided between two rear wheels and a weighted
and pivoting front wheel as wide as the vehicle (figure 6). And in 1885,
Karl Benz unveiled the Benz Patent Motorwagen, a pioneering internal-
combustion automobile that used a precursor of the steering wheel to
pivot its front wheel (figure 7). Nevertheless, the future of motorized
transport was destined to lie with four wheels rather than three.

FIGURE 6 The Aveling & Porter steamroller "Britannia," 1867.
(Photograph by BulldozerD11 / Wikipedia Commons)

FIGURE 7 The Benz Patent Motorwagen, No. 1, 1885. (Flickr.com)

Steering presents the primary drawback to placing a steam engine on a four-wheeled base. When the animals pulling a vehicle change direction in response to commands from the driver, they cause the device that connects them to the vehicle, either a central wagon beam or two shafts extending forward from the vehicle's axle or bed, to put turning pressure on the vehicle. While this kind of pressure works well with a two-wheeled cart whose wheels rotate independently of each other, it can produce only a shallow turn by wagons whose front axle is firmly fixed to the bottom of the vehicle. The rear wheels, wanting to continue forward in a straight line, prevent any sharper turn. If, however, the front axle is fixed to the bottom of the vehicle by a wooden or metal pin in its middle, it is free to pivot and change direction in response to the turning pressure exerted by the animals. The degree of the turn may still be limited if the front wheels bang into the bed of the wagon after the axle has turned only a few degrees, a limitation known as full lock, but this can be obviated either by making the front wheels small, so they can pass underneath the wagon on a sharp turn, or by elevating the wagon's bed, or sometimes just the part directly over the front axle, to achieve the same end.

Replace the animals with a motor engine, and changing direction becomes the task of the driver, who must turn the front wheels of a four-wheeled vehicle. This task demands great strength if a major portion of the vehicle's weight rests on the axle. (Anyone who has lost power steering while driving a car has experienced the sudden increase in how much force it takes to turn the steering wheel.) This is why Trevithick and Benz (though not Cugnot, who put great weight on the front wheel) opted for a single front wheel bearing a minimal load. Yet instability problems can arise even with a front axle that pivots in the middle, particularly when making very sharp turns. (Today, when steering based on a front pivot survives only on semi-trailer trucks whose horizontal circular connector, sometimes called the fifth wheel, connects the cab to the trailer, this sort of problem causes jackknifing accidents.)

The earliest publicized improvement over steering a four-wheeled vehicle by means of a pivoting front axle came from the workshop of an engineer in Munich named Georg Lankensberger in 1818. Lankensberger asked his friend Rudolph Ackermann, a German-born coach builder who had turned to lithographic printing after emigrating to England, to take out a patent on a device he had come up with. His friend complied, but put his own name on the application. Consequently, Lankensberger's invention has been called Ackermann steering down to the present day. The core of the idea was to keep the front axle stationary in rounding a curve, but pivot each wheel independently at its end (figure 8). Lankensberger's system has two advantages. First, it allows the wheels to turn at different angles so that the outer wheel can follow a longer path around a corner than the inner wheel; and second, the vehicle is less likely to become unstable and turn over on sharp curves. Later improvements fine-tuned Lankensberger's design, but turning each wheel instead of the axle between them eventually became the standard approach for automobile designers, including Benz, who abandoned his tricycle concept in the Benz Velo of 1894 (figure 9).

As a historical footnote, modern motor transport might have developed along a very different trajectory if Charles Darwin's grandfather Erasmus Darwin had pursued the design for a fixed front axle with separately turning front wheels that he had his coach builder install in two carriages around 1759. For many years, the vehicles carried the rotund physician through the countryside to see his patients, but otherwise his invention came to naught for lack of interest. Almost sixty years were to pass before Lankensberger invented (or, rather, reinvented) Ackermann steering, and it was during that half century that the railroad locomotive was developed using the wheelset technology of England's coal mines. Erasmus Darwin's membership in the Lunar Society, a small group of remarkably innovative scientific thinkers that included James Watt, who shrank Newcomen's colossus into an engine small enough to be placed on wheels, suggests how close the pioneers of steam transportation came to turning four-wheeled carriages into steam wagons instead of minecars into locomotives.

In the event, the steam locomotive evolved from the technological universe of mining, not of road transport, though ultimately the Stanley Steamers that Francis and Freelan Stanley marketed for two decades after

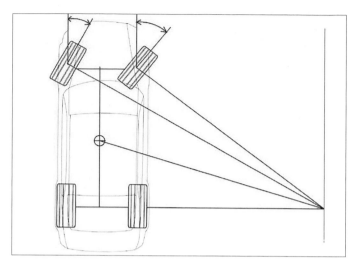

FIGURE 8 A diagram of Ackermann steering, showing the front wheels turning at slightly different angles from each other because of the radius of the curve.

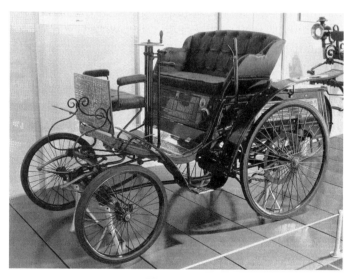

FIGURE 9 The Benz Patent Motorwagen, Velo, 1894.
(Photograph by Softeis / Wikimedia Commons)

1897 performed better than do most automobiles powered by internal-combustion engines. As part of their mining heritage, the locomotives and railcars that ushered in the railroad age used wheelsets instead of independently rotating wheels. This was possible because the tracks they traveled on, taken from mining practice, took the place of a steering mechanism. And the loads they were designed to carry consisted of heavy coal rather than comparatively light human passengers. Wheelsets were also desirable because the derailment that would be unavoidable if a single wheel broke or flew off its axle, a not uncommon phenomenon with carriages, presented a far greater threat to passengers and freight than did a flat spot that prevented a wheelset from rotating efficiently. Although railroads did eventually adopt a pivoting-wheel assembly, called a truck in the United States and a bogie in Britain, it normally consists of two wheelsets connected to each other beneath a central pivot rather than wheels rotating independently of each other on the ends of an axle.

Thus the railroad industry and the automobile industry came to be founded on competing notions of what a wheel is. What makes this important historically are the intrinsic advantages and limitations of the two different concepts. Leaving the detailed histories of trains to other historians, such as Wolfgang Schivelbusch,[9] the remainder of this chapter focuses on five areas in which the intrinsic characteristics of the competing wheel concepts shaped the modern world: road design, rolling resistance, urban planning, scheduling and velocity, and the cultural image of the driver.

Road Design

Rails were invented to make it easier to push mine-cars, which had four wheels fixed to the ends of two axles that rotated in holes going through projections underneath the cars. Although horse-powered wagon ways that used rails to carry ore outside the mines came into use in the seventeenth century, fixed wheels without rails were seldom encountered outside the mining environment except on two-wheeled oxcarts in a few rural areas. Since the axles themselves rotated, they could not be fixed to the underside of a four-wheeled wagon and thus could not take a pin in the middle to facilitate steering. Within the mines, however, mine-cars had to have four wheels, for the miners shoveling the ore into the cars could not balance the heavy loads on two wheels, nor would a two-wheeled car have stayed on a track after rails came into use. Fortunately, since the miners constructed the cars' pathways by their digging, the floors of the mines were solid stone rather than dirt and grass. Thus the miners could ensure that the cars would roll along straight corridors on floors that were both hard and flat. (Wheel ruts worn into stone pavements by repeated traffic, such as are commonly found on ancient paved roads, may have triggered the idea of installing rails to keep the mine-cars on track [figure 10].)

The first rails were of wood, but iron strips nailed to the surface of the wooden rail to protect against wear paved the way for the solid iron rails, initially only three feet long, that steam locomotives ran on. The first documented extension of a rail system outside a mine and into the open air, the two-mile-long Wollaton Wagonway designed by Hunting-don Beaumont for carrying coal out of his mines in eastern England, opened in 1604. While miners usually pushed the ore-cars inside the mines, horses took over on rail-equipped cross-country wagon ways, where a single animal could pull more than four times the weight it could manage when harnessed to a conventional wagon on a dirt or gravel road (figure 11). Whether within the mine or in the open air, the wheels of mine-cars were fixed to the axles and had flanges on the outer edge of their rims (see figure 1) that kept them on the track and prevented them from changing direction. When sufficient power was available, initially multi-horse teams but eventually steam locomotives, the cars could be linked together in trains because each of them followed the exact path of the car before it.

Laying rails and using fixed wheels dictated that the new tracked wagon ways, which became the model for steam railroads, should have only gentle slopes and gradual curves. The great weight of a train of loaded coal-cars confirmed this limitation, since going downhill at a steep angle put enormous strain on the rudimentary friction-brake systems derived

FIGURE 10 Cart ruts in stone pavement at Pompeii.
(Courtesy of Jared Lockhart, www.jaredlockhart.com)

FIGURE 11 A wagonway with a train of coal-cars pulled by horses.
(Copyright © Chronicle / Alamy)

from heavy wagons. When the earliest successful railroad, the Liverpool & Manchester, opened in 1830, it supplemented the power of the locomotive by attaching cables reeled in by stationary steam engines wherever the grade exceeded 1 foot per 100 feet. Its normal grade was 1 foot of climb for every 2000 feet of track. By comparison, modern highways are descended from carriage roads and typically go up hill and down dale with maximum grades in the area of 7 feet of climb per 100 feet of road. City streets can be even steeper. Filbert Street, the steepest in San Francisco, climbs 31.5 feet for every 100 feet of road. As for curves, even 180 years after the opening of the Liverpool & Manchester Railway, American trains normally have a turning radius of 410 feet as compared with a fairly standard 35 feet for American automobiles.

These differences meant that from the very beginning, the use of wheelsets on tracks required, in most instances, entirely new rights-of-way involving extensive cuts, fills, tunnels, and embankments. Unlike carriage roads, the forerunners of automobile highways—which commonly evolved from the paths followed for centuries by cross-country walkers, horse riders, and drovers herding animals to market—the railroad profoundly altered the landscape. Moreover, the need for constant and scrupulous track maintenance to prevent derailment required responsible ownership of each right-of-way. This vaulted railroad building to the forefront of modern industrial organization and investment. The granting of rights-of-way, following the precedent of the inland canal-building boom at the end of the eighteenth and beginning of the nineteenth century, also committed governments more deeply to the planning of railroads than had ever been the case with carriage roads.

These imperatives, dictated by the very concept of the railroad, may seem so self-evident as to be not worth singling out, but they would not have had so profound an impact on the landscape, the structure of economic enterprise, and the involvement of government in transportation if the first small steam engines had been successfully mounted on steerable road carriages, as Richard Trevithick visualized in 1803, instead of on non-steerable mine-cars. To be sure, very heavy loads would always have benefited from gentle grades and gradual curves, and steam railroads probably would have developed alongside steam cars even if the latter had been invented first. But the seventy-year gap between the construction of the Liverpool & Manchester Railway and the production by the Stanley brothers of a practical and economically competitive steam automobile gave steam-powered, fixed-wheel transport an enormous head start over motor vehicles with independently rotating wheels.

In the twentieth century, this gap was more than made up. Railroad building in industrialized countries slowed precipitately as cars and trucks surged onto the scene. The United States, for example, went from having approximately 78,000 gasoline- and electric-powered cars and 700 trucks in 1905, to 5.5 million cars and 250,000 trucks in 1918, to 254.4 million cars and 15.5 million trucks today. However, while the United States had approximately 120,000 miles of railroads in 1890, up from 9000 miles forty years earlier, as late as 1914 the country's road network outside cities and towns included only 32,000 miles of hard-surface pavement (brick, concrete, asphalt). Government support for road building did not become meaningful at a national level until 1916, when the Federal Aid Road Act offered states financial support, on a cost-sharing fifty–fifty basis, for highway improvements provided that each state establish a highway department to oversee road construction.

A new era in road construction opened with the first German autobahns in the 1930s. These high-speed highways, similar to the parkways pushed into being at the same time by Robert Moses in the New York City area and expanded nationwide in the Interstate Highway System, championed by Dwight Eisenhower and authorized by the Federal Aid Highway Act of 1956, took a page from the railroad-design book in following new rights-of-way, sharply limiting the number of access points, having merge-on / merge-off intersections to obviate stop-and-go driving, and incorporating gradual curves and grades. They differed, however, in paying greater attention to the landscape, restricting commercial advertising, and catering to private passenger vehicles. As Fritz Todt, the inspector general for German roadways, described the autobahn project in 1934:

The routing has to be designed differently for the automobile than for the railroad. The railroad is a medium for mass transit (also for masses of people). The roadway is a medium for individual transport. The train is usually a foreign body in the landscape. The motor roadway is and remains a road, and the road is part and parcel of the landscape. The German landscape is full of character. The motor roadway, too, must be given German character.[10]

At the same time that Todt was overseeing the autobahns, Frank Lloyd Wright was envisioning the futuristic Broadacre City, in which there might be one railway station but where most people would drive in private automobiles to the shopping mall from their suburban homes on one-acre lots, using service roads to get to merge-on / merge-off access points on a divided, limited-access highway (figure 12).

Rolling Resistance

The second area in which vehicles with wheelsets differ significantly from those with independently rotating wheels is rolling resistance. Steel railcar wheels rolling on steel rails move loads with far greater efficiency than almost any other combination of wheel and surface. The resistance of any wheels to rolling—that is, the degree to which contact with the surface, whether rail or road, slows the movement of the wheel—is measured by the coefficient of rolling resistance. This measurement plays a major role in determining how much energy it takes to set a vehicle in motion and how long it will remain in motion on a flat plane with its power shut off.

A railroad car has a rolling resistance that is only half that of a mine-car, even though both use steel wheels and steel rails. This is because railroad wheels have a much greater diameter than mine-car wheels. But both types of fixed-wheel cars have much less rolling resistance than vehicles with independently rotating wheels. It takes almost ten times as much energy to set an automobile with rubber tires in motion on a concrete pavement as it takes to start a railroad car. And a nineteenth-

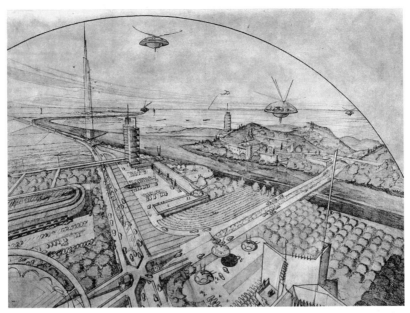

FIGURE 12 Frank Lloyd Wright, "Broadacre City: The Living City" (1958). (mediaarchitecture.at)

century stagecoach on a dirt road required over four times as much energy as an automobile does, and twelve to twenty times as much as a mine-car. These great discrepancies reflect both the traction with the road that carriages and automobiles require when starting, stopping, and negotiating curves, and the very slow development of smooth and durable road surfaces.

John McAdam, a Scotsman born in 1756, introduced a new and superior approach to building carriage roads, but controversy over its value persisted through much of the nineteenth century. The Roman road system of ancient times had long provided the conceptual framework for European road construction. Roman engineers laid large flat stones on top of a layer of cement that was itself underlain by layers of small loose stones. Roads with this design served their primary function of moving marching legionnaires around the empire quite well, but heavy vehicles with iron-rimmed wheels broke and rutted the surface stones so badly that the law code of Emperor Theodosius, promulgated in 438 C.E., set load limits—145 pounds for a light two-wheeled cart, and 1089 pounds for the heaviest wagon—at levels far lower than the actual carrying capacity of those vehicles.

By the time of the carriage revolution in the sixteenth century, road builders had reversed the Roman system and were putting down large stones or paving blocks first as the bottom layer and then covering them with small stones, often gravel from streambeds. These roads held up somewhat better to wheeled traffic, but they were expensive to build and maintain, and the responsibility for maintenance was frequently unclear. McAdam's innovation was to use only small stones, but small stones with sharp edges, as opposed to tiny stones or rounded pebbles from streambeds. If the size was right—building supervisors sometimes put them in their mouth to check—the coach traffic would cause the stones to press together more tightly and fill in the spaces between them instead of being pushed to the shoulder by passing wheels, as happened with rounded gravel. McAdam's insistence that the best way to guarantee small size and sharp edges was to have workers, including women and children, sit beside the road and use hammers to break up larger rocks led eventually to the American cartoon image of convicts "working on the rock pile" to produce stones for the "chain gang" to use in road repairs (figure 13).

The qualities that recommended small, sharp-edged stones for road construction benefited the railroad industry as well. The crossties (in England, sleepers) that keep railroad tracks, as opposed to mine tracks, at a constant gauge, or distance from each other, rested on leveled ground in the earliest stages of railroad building; but it was soon found that laying the ties on a ridge of McAdam's stones, called ballast, provided improved solidity and better drainage (figure 14).

The word "macadam" entered road-building parlance after 1820 as the new approach to surfacing spread. It did not make the problems of

FIGURE 13 A convict working on the rock pile.
Cartoon by Jack Corbett / cartoonstock.com)

FIGURE 14 Stone ballast underneath railroad crossties.
(Photograph by BesigedB / Wikimedia Commons)

dust, mud, and deterioration from stones being crushed or pushed aside by passing wheels go away, but cross-country travel did become somewhat less taxing. As for paving experiments using wooden blocks, bricks, asphalt (a mixture of bitumen and small stones), and cement that became more frequent toward the end of the century, they remained confined mostly to city streets until after World War I.

Paving improvements reduced but did not eliminate the gap between the rolling resistance of railroad trains and that of carriages and automobiles. To this day, the cost of carrying one ton of cargo for one mile by rail is several times lower than the cost of carrying the same load in a truck. However, even though the human beings in a passenger coach weigh almost nothing compared with a seventy-ton carload of coal, an Amtrak Viewliner car is almost double the weight (fifty-eight tons) of an empty hopper car (twenty-seven tons).

Meanwhile, driving a car became cheaper and cheaper as road surfaces improved and the rolling resistance of passenger vehicles diminished. As the economic efficiency of driving an automobile, not to speak of its speed and convenience, came closer and closer to the energy cost per ton-mile of carrying people by rail, passenger trains went into eclipse, even though freight trains continued to provide a superior form of bulk hauling. In Europe, high taxes were placed on gasoline in the post–World War II decades in order to reduce the pressure to build improved roads and redesign cities. This slowed, but did not prevent, a shrinkage of passenger-rail service, at least until the advent of high-speed Japanese bullet trains (Shinkansen) in 1964.

On a global basis, whether a country developed a railway network or a road network depended on when motor transport began in relation to the shifting balance of efficiency between railroads and automobiles. British India acquired an extensive railroad system in the mid-nineteenth century at a point when cross-country roads were mostly dirt tracks; today, India is twelfth in the world in miles traveled by rail per person per year, right after Japan and eleven European and post-Soviet countries. What railroads there are in many sub-Saharan African countries, though, were built after the advent of the automobile and were designed to carry commodities or mine products rather than people. So highways were built for passenger traffic, with the result that today fifteen sub-Saharan countries rank above India in per capita ownership of motor vehicles.

Urban Planning

Wheeled vehicles and cities have always had an antagonistic relationship. The circulation patterns of non-vehicular cities, including many historic city centers in the Middle East, often incorporate staircases, sharp corners, very narrow lanes, and even ladders, since people on foot have no difficulty negotiating such obstacles. Porters with saddles on their backs can still be found in pedestrian-scale locales like Istanbul's covered bazaar (figure 15).

FIGURE 15 A human porter carrying loads on a pack saddle in Turkey.
(© Melvyn Longhurst / Alamy)

This situation changed when wheeled vehicles entered the picture. Checkerboard grid designs with straight streets and right-angle intersections appeared from Rome to China in ancient times. In post-Roman Europe, however, urban street grids deteriorated, and carts and wagons were commonly unloaded at city gates, with their goods being distributed within the city by porters or pack animals. Aside from Chinese wheelbarrows, visual evidence for hand-pushed vehicles in medieval cities appears only rarely, presumably because of rough or nonexistent paving.

The carriage revolution of the sixteenth century ushered in a period of change during which the streets of European cities slowly acquired better paving, straighter alignment, greater width, and freedom from impediments. The absence of similar changes in non-Western lands, where difficult-to-steer four-wheeled vehicles were rare or unknown, led European travelers to represent the cities they visited there as having narrow, dark, and pestilential mazes of streets and lanes that were out of step with the modern world. In fact, pedestrian-scale living is very supportive of social and personal relationships, since a street or lane can bring neighbors together instead of keeping them apart.

With carriages, freight wagons, and improved paving came horse manure, traffic congestion, and a heightened danger of being run over. Although the first of these nuisances disappeared with the spread of the automobile, the other two steadily worsened. Pedestrians, for whom city streets had originated, became confined to sidewalks and exercised ever-greater caution in crossing thoroughfares designed for motor vehicles. One can easily imagine the astonishment that someone born before 1600 would express in discovering that people today voluntarily live in blocks of buildings cut off from other blocks of buildings by wide streets reserved for huge, fast, and dangerous machines. The city street has become a desert.

Of course, these developments involved vehicles with independently rotating wheels and began well before the advent of the steam engine. Fixed-wheel railroads involved a different set of constraints. In the name of safety and in keeping with private ownership of rights-of-way, fences and embankments barred pedestrian access to train tracks within the city, and crossing barriers gave trains priority over both pedestrians and rotating-wheel vehicles. Surveyors laying out straight and level tracks necessarily ignored existing street networks, with the result that railroads often split communities in two. Hence the proverbial distinction between the "wrong" and the "right" sides of the tracks. Railroad stations that arose as magnificent expressions of urban pride commonly bordered on urban dead spots in the form of the vast areas needed to store railroad cars and assemble them into trains.

Streetcar lines mitigated the divisive impact of fixed-wheel transport on urban life. Light-rail vehicles, as they are now called, grew out of horse-car rail systems and operated alongside carriages and automobiles. They made many stops instead of converging on single terminals, and their tracks, which were usually set into the pavement, did not obstruct pedestrian movement to the same degree as railroad tracks. Moreover, because streetcars were much lighter than train cars and their short routes permitted passengers to stand as well as sit, the transport cost per person was much lower than that of a passenger train. This efficiency contributed to their being instrumental in fostering suburbs and creating a culture of commuters who lived some distance from the parts of the city in which they worked. Subways combined features of railroads and streetcars. They had few stations and did not have to compete with automobile traffic, but they used lighter cars and focused exclusively on passengers.

Many streetcar systems disappeared in parallel with the spread of the private automobile, particularly in the United States. Although the reasons for the abandonment of this effective form of mass transit varied from city to city, one underlying factor was the fact that streetcars did not carry freight. Thus the financial returns that railroads continued to profit from even in the face of declining passenger traffic were not available to buffer the streetcar industry against the changing transport preferences of its customers.

Light rail came and went, and in some places is coming back. Monumental railroad stations came and went, though a few are now being salvaged. Railyards came, underwent abandonment or semi-abandonment, and are now targets for urban redevelopment. The urban sun has not fully set, however, on the wheelset-on-rails concept. Although subway systems in Europe and North America expand with difficulty because of the enormous expense of construction, an enthusiasm for underground railroads continues to spread in other parts of the world. Subways may not offer the convenience of frequent stops, but like high gasoline taxes, they help constrain the spread of passenger cars onto street networks that in many non-Western countries are still better suited to pedestrians than to automobiles.

Scheduling and Velocity

In 1840, the Great Western Railway in England instituted a precise synchronization of clocks that it called railway time. Thus began what slowly became a worldwide transition from local times, which might vary by several minutes from town to town, to times that were standardized within geographically fixed time zones. A quite realistic fear of train collisions, particularly on railroads that did not have double tracks to accommodate trains traveling in opposite directions, prompted this innovation. Handheld timepieces had reached the necessary level of precision during the eighteenth century; but apart from determining longitude at sea, precise synchronicity between geographically separated locations had never before been required. Eventually, of course, all governments, industries, and communication systems switched to standardized time and took it for granted in their activities, but railroads set the trend in motion, and an accurate pocket watch became the hallmark of a trainman.

Along with precise timekeeping came schedules by which travelers and commuters could see in advance exactly how their journey would progress. This did not mean that trains always arrived on time, but it created that expectation. In 1926, for example, the New York Central Railroad offered passengers a $1 refund (enough for one and a half gallons of milk) for every fifteen minutes its prestige train, the Twentieth Century Limited, was late on the run between New York City and Chicago. Timetables also appeared for vehicles using independently rotating wheels, like buses and trucks, not to mention ferries and airplanes; but none of these modes of transport had quite so crucial a need for precise timing.

Railroad timetables assumed that trains would travel at predicted speeds, and this sparked a previously unknown exploration of the concept of velocity. The speed of a horse-drawn vehicle or of a ship had always excited interest, of course, and racing had drawn enthusiastic spectators from antiquity onward. However, the precise speed at which a winning charioteer navigated a racing oval or a winning sailboat rounded a buoy made no difference as long as one chariot or sailboat beat out the others in the competition.

This all changed with the advent of motorized vehicles, first trains and then automobiles. Since an engine provides a controllable amount of power that is unaffected by the muscles and stamina of a horse or the vagaries of the wind, it became possible not only to predict the speed at which a train would run, and hence the exact time of arrival noted on its timetable, but also to design locomotives that would run faster and faster in absolute terms, as opposed to competing with another locomotive racing alongside them. The first steam locomotives, around 1814, traveled at 4 miles per hour. By 1832, the speed of top locomotives reached 60 miles per hour, and by 1938 a peak for steam-powered trains was reached at 126 miles per hour. The speed records for a train with a diesel engine, 148 miles per hour, and one powered by electricity, 357 miles per hour, were set in 1987 and 2007, respectively.

The Victorian pioneer of digital computing, Charles Babbage, facilitated this fascination with velocity by inventing a speedometer for locomotives. Both the Roman Empire and the contemporary Han dynasty in China had utilized mechanical odometers for measuring distances traveled, and they may have existed as far back as the time of Alexander the Great, but Babbage's device seems to have been the first to measure absolute speed. Not until 1886 and the dawn of the automobile age, however, did a speedometer designed for carriage and automobile wheels make an appearance, the brainchild of a Croatian inventor named Josip Belušić.

Speed attracted even more attention in the twentieth century. In the vehicular realm, automobile racing got going almost as soon as there were cars to race. The late-nineteenth-century craze for bicycling probably encouraged a racing mentality in those early automobile experimenters who began as bicycle mechanics, for the earliest bicycle race took place in the Parc de Saint-Cloud in Paris in 1868, when motorcars were still in the embryonic stage; and bicycle racing was so well established by century's end that it was included in the first modern Olympic Games in 1896. The Latin motto for the competitions, *Citius, altius, fortius* (Faster, higher, stronger), proposed by Pierre de Coubertin, who founded the modern games, enshrined measurable speed as a core civilizational value; and the stopwatch, developed in the early nineteenth century for astronomical observations such as tracking the passage of Jupiter's moons across the planet's face, took its place as a tool of modern culture alongside the speedometer.

The commonplace sentiments that change happens faster in modern times, that people today move and communicate faster than ever before, and that history itself has accelerated betray the degree to which the idea of speed, as a positive value, has permeated Western society. For many people, "life in the fast lane" epitomizes being on top of society's fast-moving changes, but there would be no fast lane were it not for the competition between rails and roads that marked the nineteenth century.

Cultural Image of the Driver

As later chapters make clear, a change in men's attitudes toward riding in wheeled vehicles contributed mightily to the carriage revolution in Europe. Carriages replaced warhorses, armor, and painted shields as emblems of noble rank, and a rising non-aristocratic middle class with no claim on martial status could now flaunt the same symbols. Carriage owners did not initially drive their own vehicles, however. Coachmen and postilions, the men who rode on guide horses in multi-horse teams, were essentially servants until cross-country stagecoaches and urban taxis made carriage driving a trade. Although some noblemen, like the father

FIGURE 16 Henri de Toulouse-Lautrec, *Alphonse de Toulouse-Lautrec Driving His Four-in-Hand* (1880). (Musée Toulouse-Lautrec, Albi, France. Scala / White Images / Art Resource, NY)

of the artist Henri de Toulouse-Lautrec, mastered the art of four-in-hand driving, which entails controlling four horses at once (figure 16), most carriage owners chose not to drive themselves until roads improved in the nineteenth century and a variety of lightweight vehicles like gigs, phaetons, and one-horse shays gained popularity.

Surprisingly, given their earlier history as passive riders in vehicles controlled by men, European and American women took to driving light carriages as well (figure 17), which raises the question of how the negative stereotype of the woman automobile driver arose. To be sure, hand cranking an early automobile could be both difficult and dangerous, but Charles F. Kettering solved this problem when he invented the electric

FIGURE 17 Arthur J. C. Rolfe, *A Smart Turn-out* (1902). (From *Penrose's Pictorial Annual, 1902–1903: An Illustrated Review of the Graphic Arts*, vol. 8, ed. William Gamble [London: Penrose, 1902–1903]. HIP / Art Resource, NY)

starter motor in 1911. In any case, controlling a horse could pose challenges for drivers of either sex if something startled the animal or traffic was badly congested. Yet cautions about women drivers, sometimes backed up by pseudoscientific assertions purporting to demonstrate female technical or emotional limitations, seem to have arisen quite soon after automobiles began to be seen on city streets. This serves as a reminder that the often ridiculed ban on women driving in Saudi Arabia, which Muslim zealots maintain honors a tradition from the Prophet Muhammad's time of men leading women's camels, may have sounded quite reasonable to American men in 1920. Yet the automobile industry did not support gender bias. Advertisements going back to the World War I era show women happily driving (figure 18), though American ads focusing on family transportation normally show Dad at the wheel.

One area, however, emerged as an almost exclusively male preserve: automobile racing. Aside from the Twenty-four Hours of Le Mans, a race in which women participated as early as 1930, it took a half century and more for women to be allowed to compete in most major races: 1949 for the top NASCAR series, 1958 for Formula One, 1976 for the Indianapolis 500, and 1977 for the Daytona 500. No woman reached a racing pinnacle until current superstar Danica Patrick was named Indianapolis 500 Rookie of the Year in 2005. Not only is the world racing fraternity composed almost entirely of men, but it has historically recruited very few drivers from East Asia, South Asia, the Middle East, and Africa. Many factors doubtless contribute to the gender and ethnic skewing of the racing universe. But as will become apparent later in this book, the five-thousand-year history of wheels in Indo-European societies—specifically in Europe, including its former colonies, and North America—testifies to an affinity between vehicle driving and male identity in cultures that descended from the Proto-Indo-European linguistic tradition. Since the earliest days of wagon nomads and chariots, through the carriage revolution of the sixteenth century, and down to the automobile era, men brought up in European (and Euro-American) societies have repeatedly linked their manhood to their vehicles.

FIGURE 18 A woman driving a Studebaker, advertisement for Big-Six model, 1920.
(Photograph by Carlylehold / Flickr.com)

This yen for driving probably contributed to the rapidity with which passenger cars overtook train and streetcar travel in popularity. Railroads produced few folk heroes: John Henry, who died after victoriously pitting his hammer against a steam drill in driving spikes into crossties; Casey Jones, who died while blowing his steam whistle to warn people on a stalled train ahead that he could not stop in time to avoid colliding with it; and Steve Broady, who died while trying to get the Southern Railway's Fast Mail—"Old 97"—to Spencer, North Carolina, on schedule:

Well they gave him his orders in Monroe Virginia
Sayin' Steve you're way behind time
This is not 38, this is Ole' 97
You must put her into Spencer on time

. . .

It's a mighty rough road from Lynchburg to Danville
It's a line on a three-mile grade
It was on that grade that he lost his airbrakes
You can see what a jump he made

He was goin' down that grade makin' 90 miles an hour
When his whistle broke into a scream
He was found in the wreck with his hand on the throttle
Scalded to death by the steam

That's about it for fixed-wheel adventure . . . leaving out train robbers and terrorists. No matter how much the steam-era passenger may have appreciated the speed and comfort of a train trip, the passivity of just sitting and enduring the clackety-clack and the windows that had to be kept closed to keep out the coal smoke contrasted badly with the freedom of automobile travel. Writers of action movies almost always portray their heroes on top of or clinging to the side of a moving train, not sitting inside or even driving it.

By contrast, the gazillion or so car chases that have flooded the silver screen and television sets over the years almost always feature male drivers, which simply confirms the excitement gap between wheelsets on rails and independently rotating wheels, and the heroism gap between male and female drivers. Screenwriters have put thousands of macho men behind the wheel, but they have given us only one Thelma and Louise.

Whether the railroad industry would ever have come into existence without a prehistory of wheelset-equipped mine-cars guided by rails lies in the realm of conjecture. Similarly, one can question whether the automobile would ever have been invented, or whether it would have been invented differently, if, as in the rest of the world, road transport in Europe had relied on two-wheeled carts rather than four-wheeled wagons and carriages. Competing concepts of the wheel—the wheelset, in which wheels and axle rotate together, and the axle with independently rotating wheels—shaped the transportation history of the industrial age. If either concept had dominated, or if only one of the two had come into play, the landscape and economy of the modern world would be very different. Surprisingly, the same thing might be said of the fourth millennium B.C.E., when both types of wheel first came into use. But not everyone was impressed with wheels.

two

Why Invent the Wheel?

The peoples of North and South America did not use wheels for transportation before the voyages of Columbus. They carried loads on their backs and on travois, splayed pairs of sticks dragged along the ground by harnessed dogs. In the Andes, they also used pack llamas. Many students of pre-Columbian society, including anthropologists, archaeologists, and historians, have expressed puzzlement over the absence of wheeled vehicles. How, they ask, could civilizations that built magnificent cities and cultivated the arts and sciences have failed to invent something as invaluable as the wheel? Particularly when they knew how to make wheeled toys.

Nineteen hundred years ago, a potter in southern Mexico takes a lump of pale clay and shapes it into a dog for her daughter to play with. Spread legs, lolling tongue, curling tail, big head with jug-handle ears. She pokes two sticks crosswise through the animal's paws, working them about until they turn easily in the holes she has made. Then she fashions the wheels. She rolls out four clay disks one inch or so in diameter and pokes holes through them. After she bakes the dog and the disks in her kiln, she pushes the sticks back through the holes in the paws and

FIGURE 19 A pre-Columbian toy dog on wheels.

attaches the wheels to their ends. Voilà. A toy on wheels (figure 19). Her daughter loves it.

Nineteen hundred years pass. An archaeologist finds the toy dog, replaces the lost originals with new sticks, and expresses puzzlement that the Olmec people dwelling in Tres Zapotes had the idea of the wheel but never built a cart or wagon. They lived in cities, fashioned giant stone heads, and had rudimentary writing and calendrical systems. So how, he asks, could a people who clearly understood the concept of the wheel have failed to build a usable wheeled vehicle?

From the potter's point of view, four clay disks attached to sticks going through the paws of a toy dog may simply have been a wonderful solution to the problem of how to entertain her child. But modern inquirers who share the belief that the wheel is the world's greatest invention have a hard time grasping that down to modern times, many, if not most, of the world's peoples did not avail themselves of wheeled transport even when they knew about it.

The most common reason advanced to explain the absence of carts and wagons in pre-Columbian America is that domesticated animals are needed to pull wheeled vehicles, and the Western Hemisphere was devoid of large animals suitable for domestication. This explanation makes no sense, however. The celebrated world historian Jared Diamond, one of the most frequently cited proponents of this view, puts the case this way: "Ancient Native Mexicans invented wheeled vehicles with axles for use as toys, but not for transport. That seems incredible to us, until we reflect that ancient Mexicans lacked domestic animals to hitch to their wheeled vehicles, which therefore offered no advantage over human porters."[1] What Diamond fails to grasp is that wheeled vehicles need not be pulled by animals. People can push or pull them just as easily as people can carry packs on their backs. The value of a wheeled conveyance comes from the weight of its load being supported by wheels instead of by human muscle and from the reduced amount of friction involved in rolling something, as opposed to dragging it along the ground, as on a travois or a sledge (a sled with runners used to traverse snowless ground [figure 20]).

Today we opt for wheels over backpacks every time we push a grocery cart down a supermarket aisle or pull a piece of luggage through an airport. It never occurs to us that we might use an animal instead of our own muscles. To be sure, the floors of supermarkets and airports are smoother than were the streets of pre-Columbian Mexico, but there is ample pictorial evidence that humans have pulled and pushed wheeled vehicles throughout our history.

Animal power is good, but it is not absolutely necessary. And it certainly was not necessary in the many societies of the Western Hemisphere that relied on the muscle power of thousands of slaves and bonded laborers. If Columbus did not find any wheels when he arrived in the New World, it was because the indigenous peoples did not find them useful, not because they lacked domestic animals. To turn this statement around, the reason that oxcarts did exist in the Spain from which Columbus departed is that some five thousand years before he was born, copper

FIGURE 20 A print after Austen Henry Layard's drawing of an
Assyrian bas relief that shows men pulling carts and moving a giant stone sculpture
on skids, ca. 700 B.C.E. (From Jessie Noakes, *Art History and Literature Illustrations*
[London: Virtue, ca. 1900–1920]. HIP / Art Resource, NY)

miners in eastern Europe decided that wheels *were* a good idea. And
those people pushed their vehicles by hand. Animal power came later.

The Olmec of southern Mexico was not the only civilization to be
unimpressed by the idea of the wheel. Most people felt the same. Cattle-
herding societies stretching from Senegal to Kenya across the Saharan
Sahel, the great desert's southern borderlands, had vast herds of domestic animals, but no wheels. It wasn't that they did not know about the
wheel. Ancient rock paintings in the southern Sahara preserve scores
of images of horse-drawn two-wheeled chariots, probably painted by
peoples who moved into the region from the north as the cattle herders responded to increasingly arid conditions by moving south. Nor was
it a question of unsuitable terrain. The Sahel is generally flat, dry, and
sparsely wooded. Moreover, the tse-tse fly belt, which makes equatorial

WHY INVENT THE WHEEL?

Africa dangerous for domestic livestock because of the parasitic diseases that the flies spread, begins farther to the south.

Cultural practices from the Mediterranean region could, of course, have reached sub-Saharan Africa from the north by way of the Nile Valley, and many did. But wheeled vehicles were not known in Egypt in the age of the pyramids. Although the ancient Egyptians had potter's wheels and were in trading contact with Mesopotamia (the valley of the Tigris and Euphrates Rivers), where oxcarts and battle wagons came into use as early as 3000 B.C.E., the technology of wheeled transport did not appear in the Nile Valley until the pharaohs adopted the war chariot around 1600 B.C.E. That the early Egyptians knew about the wheeled vehicles in use in Mesopotamia, but chose not to adopt them, cannot be doubted.

The Egyptians' disinclination to build carts and wagons sheds skeptical light on the notion that the wheel was originally understood to be a wonderful idea. David W. Anthony, the most recent scholar to explore the idea in his excellent book *The Horse, the Wheel, and Language*, puts the case this way:

> It would be difficult to exaggerate the social and economic importance of the first wheeled transport. Before wheeled vehicles were invented, really heavy things could be moved efficiently only on water, using barges or rafts, or by organizing a large hauling group on land [see figure 20]. Some of the heavier items that prehistoric, temperate European farmers had to haul across land all the time included harvested grain crops, hay crops, manure for fertilizer, firewood, building lumber, clay for pottery making, hides and leather, and people . . . *the introduction of wagons passed on the burden of hauling to animals and machines, where it has remained ever since.*[2]

This sounds very sensible. Indeed, from this point of view, wheeled transport seems as obviously desirable as sliced bread or the cell phone and therefore must have spread very rapidly. But even though Anthony is a superb archaeologist and historian, his statement of the case does not jibe with historical reality. It isn't just a matter of the Olmecs, the Egyptians, and the peoples of the Sahel unaccountably deciding to move their firewood, hides, and manure without using wheels. People and animals had been hauling those loads on their backs for many thousands of years before the wheel was invented, and in most parts of the world they continued to do so for several thousand years more before motorized vehicles took over in the twentieth century.

The items on Anthony's list of things that needed hauling share a feature with virtually everything that was commonly transported in early societies. They were divisible into loads of manageable size—that is, loads that could be placed on the back of an ox, a donkey, a horse, or a camel . . . not to mention a man or a woman. This holds true even for the "temperate European farmers" whom Anthony singles out. Visit any art museum and look at European landscape paintings. Before the late eighteenth century, rural scenes feature far more pack and riding animals than they do carts, wagons, and carriages. Pack and riding animals did not need paved roads, they could step over or around obstacles, they required little direction, and their saddles were cheap and durable. Instead of making wheeled transport inevitable, as Diamond proposes, domestic animals provided the best means of moving goods overland without wheels. Unlike pack and riding animals, carts, wagons, and carriages fared miserably on rutted and muddy roads: rocks and fallen tree limbs had to be cleared away for them to pass, the driver had to sit on a bouncing seat and maintain constant control of the animals, and both wheels and axles were prone to breakage.

Conditions outside Europe differed little from those within, except for the general absence elsewhere of four-wheeled vehicles. In the Middle East, where two-wheeled carts and four-wheeled battle wagons came into use in the third millennium B.C.E., wheeled transport all but disappeared during the first five centuries C.E. This despite the fact that aridity kept the roads in the region clear of mud and snow during most of the year, and ancient deforestation relieved road builders of the need to fell trees. The reason for the abandonment of the wheel from Tunisia to the eastern frontier of Iran (but not from the Anatolian heartland of modern Turkey) had much to do with the realization that most loads could be carried more cheaply by pack camels and donkeys than by oxcart as long as the region's population included a substantial number of pastoral nomads engaged in supplying livestock and conducting caravans. Since few caravan routes passed through sandy wastes, the unique suitability of camels to that terrain was much less important than their great load-carrying capacity.

My book *The Camel and the Wheel* presents evidence of the region's turning away from the wheel use that is so evident in ancient pictorial remains and in texts like the Bible.[3] But the totality of the abandonment comes through best in eyewitness accounts, such as that published in 1833 by the Englishman John Kitto, who wrote under the pen name Deaf Traveller, describing his southward journey from the Caucasus Mountain area west of the Caspian Sea into northern Iran:

> In the journey to Bagdad we had traveled in English landaus from Petersburgh to Teflis, where, leaving them to be sold, we proceeded to Shausha, in the Karabaugh, in wagons, without springs, belonging to the German colonists in Georgia: the roads then becoming impracticable to wheel-carriages, we were obliged to perform the rest of the journey on horseback in Persian saddles. . . . I saw no wheel-carriages of any kind in Persia.[4]

Farther to the east, a comparison between China and Japan highlights the option that different societies had in whether to adopt or not adopt the wheel. Tradition maintains that China borrowed the use of the war chariot from the nomadic peoples of the Eurasian steppes during the reign of the emperor Wu Ding of the Shang dynasty, who lived around 1200 B.C.E. But mounted cavalry rapidly displaced chariots on the battlefield after 300 B.C.E. Chariots for use by noblemen faded out more slowly, but by the twelfth century, when the famous artist Zhang Zeduan executed a magnificent and immensely detailed scroll painting showing street life in a Chinese city, the chariots were gone.

The most common wheeled vehicle in Zhang Zeduan's painting, and in an even larger scroll painting from the eighteenth century conceived as an updating of his masterpiece, is the wheelbarrow. Each wheelbarrow has a single large wheel and balances its load, either bundles or two passengers sitting on either side of the wheel, directly over the wheel's axle (figure 21). Thus the Chinese wheelbarrow differs greatly from the European wheelbarrow, a later and apparently independent invention, in that the operator does little lifting and devotes his strength instead to stabilizing the load and pushing the vehicle forward. The wheelbarrows we use, descended from medieval European designs, place the load between the wheel and the handles and thus require the operator to simultaneously lift and push. Although an animal or a second man is occasionally shown pulling a Chinese wheelbarrow that is simultaneously being stabilized and pushed from behind, animal power serves simply as a complement to human power. Wind power may also supplement muscle power by means of a small mast and sail rising over the single wheel. In general, both scroll paintings show two human porters and pack or riding animals for every cart and wheelbarrow. And members of the elite travel in enclosed sedan chairs or palanquins (figure 22).

Traffic in premodern Japan looked very different from that in China because the Japanese used wheeled vehicles in only limited ways. This despite the fact that China and Korea strongly influenced the development of Japanese culture. Horse riding, a practice borrowed from Korea, shows up early in Japan in the form of ceramic figurines of horses found in Japanese burials dating between the third and the sixth century. So there is no questioning the presence in Japan of strong domestic animals, and it is inconceivable that the Japanese were unaware of the use of horses to pull carts and chariots on the Asian mainland. Yet down to the mid-nineteenth century, horses were seldom or never used to pull vehicles. As for oxen, which the Japanese used to pull plows, they too were rarely harnessed to carts and never to four-wheeled wagons.

Here is how an anthropologist and specialist on Japanese rural life, Alan Macfarlane, summarizes the evidence:

> In the middle of the nineteenth century, a number of western commentators noticed the absence [of animal-drawn wheeled transport]. [Lawrence] Oliphant [in country in 1861] on the Elgin mission noted in almost identical words to [Engelbert] Kaempfer [1690–1692] and [Carl]

FIGURE 21 Chinese wheelbarrows with passengers, 1910–1925. (Library of Congress, Prints and Photographs Division / Frank and Frances Carpenter Collection)

FIGURE 22 Zhang Zeduan, *Along the River During the Qingming Festival* (twelfth century). (Collection of the Palace Museum, Beijing. Copyright © The Metropolitan Museum of Art, New York)

Thunberg [1775–1776], "I also observed, for the first time, one or two carts of a very rude construction, and drawn by bullocks; but they are apparently very little used in Japan." [Rutherford] Alcock [1858–1864] quoted [John] Veitch [d. 1870]: "There are no carts in this district. Everything is transported from and into the interior by horses and bullocks." [Edward] Morse [1877–1880] noted that "I have seen no wheeled vehicles except the [hand-pulled] *jinrikisha* and there are very few of these."[5]

Visual evidence corroborates these observations and indicates the effectiveness of the banning of wheels on highways by the Tokugawa shoguns, who ruled Japan from 1600 to 1868. In 1833 and 1834, the great Japanese artist Utagawa Hiroshige published a series of woodblock prints, *The Fifty-three Stations of the Tōkaidō*, illustrating sites along Japan's main coastal highway between the major cities of Edo (Tokyo) and Kyoto. With a colleague, Keisai Eisen, he subsequently produced another series, *The Sixty-nine Stations of the Kisokaido*, showing scenes along the inland highway that connected the same two cities. Hundreds of wayfarers and city folk appear in these prints: people walking, riding horses, being borne in palanquins suspended from poles carried on the shoulders of bearers, and carrying heavy cargo supported in the same fashion (figure 23). But in only one place, Otsu, a seaport town that marked the southern end of both roads, do heavily laden oxcarts appear (figure 24). Moreover, photographs of the original paving on the few surviving sections of these major roads show such rough and irregular stonework that no wheeled vehicle could possibly have passed over it.

Wheeled vehicles became more common as Japan modernized in the late nineteenth century, but those pulled by humans predominated. In the cities, this expansion of wheeled transport took the form of rickshaws (chapter 10). But an Englishman who lived in Japan during World War I observed another new form, hand-drawn carts to carry human waste from latrines in town to the countryside to be used as fertilizer:

As is well known, hardly any of the night soil of Japan is wasted. Japanese agriculture depends upon it. Formerly the night soil was removed from

FIGURE 23 Utagawa Hiroshige, *Kusatsu: Famous Post House*, a woodcut
showing potters at the fifty-second station of the Tōkaidō.
(From *Fifty-three Stations of the Tōkaidō* [1833–1834]. Art Resource, NY)

FIGURE 24 Utagawa Hiroshige, *Spring Water Teahouse*, a woodcut
showing oxcarts at Otsu, the fifty-third station of the Tōkaidō.
(From *Fifty-three Stations of the Tōkaidō* [1833–1834].
Newark Museum, Newark, N.J. /Art Resource, NY)

the houses after being emptied into a pair of tubs which the peasant carried from a yoke. Such yoke-carried tubs are still seen, but are chiefly employed in carrying the substance to the paddies. The tubs which are taken to dwellings are now mostly borne on light two-wheeled handcarts which carry sometimes four and sometimes six. A farmer will push or pull his manure cart from a town ten or twelve miles off. It is difficult to leave or enter a town without meeting strings of manure carts.[6]

Another distinctive use of wheels in Japan is on elaborate floats called *dashi*, which feature in festivals in many parts of the country. These vehicles can be two or three stories high and have one, two, or three wheels; but none of the axles pivot to allow the float to be steered. *Dashis*, like rickshaws and farmers' fertilizer carts, are normally propelled not by animals but by men pulling on beams and ropes and pushing from behind (figure 25).

FIGURE 25 A Japanese *dashi* being pushed at a festival.
(Copyright © Malcolm Fairman / Alamy)

WHY INVENT THE WHEEL?

Contrary to Diamond's assumption that if suitable domestic animals are present, simply thinking about wheels will inevitably lead to the development of carts and wagons, the case of Japan speaks eloquently to the uncertainty of any society's use of the wheel. To be sure, Japan is mountainous and forested, and, as Hiroshige's woodblock prints show, its traditional cross-country routes involved fords over many streams and rivers. But the same observations can be made about much of premodern Europe. In either region, carts could have been used in cities and towns even if there were limitations on their practicality in the countryside.

So why did the Japanese—like the pre-Columbian Olmecs, the Arabs (after 600 C.E.), and the African inhabitants of the Sahel—prefer to walk, ride horses or donkeys, drive pack animals, or carry goods on their back instead of using wheels? There is no simple way of knowing, particularly if the society in question included craft workers who commonly used wheels or other rotary devices for throwing pots, spinning thread, reeling yarn, grinding grain . . . and making toy dogs.

But perhaps we are asking the wrong question. Rather than puzzling over what we imagine to be a failure to recognize a great invention, it makes more sense to assume that alternative modes of transportation fully satisfied the needs of those peoples who knew of wheels but chose not to use them, just as they did worldwide for tens of thousands of years before the invention of the wheel. Operating on that assumption, the question to ask is: Where, when, and why did a transportation need arise that could not readily be met by the traditional technique of dividing a load into manageable sizes? Even if wheeled transport soon spread to other regions once it was invented, and in its expansion was adapted to purposes other than the one that inspired its invention, its point of origin should exhibit some particularity in terms of how the first wheels were used.

three

A Square Peg in a Round Wheel

This chapter advances a new theory. It proposes that the wheel was invented for use in copper mines in the Carpathian Mountains of eastern Europe and that four-wheeled mine-cars in that region were pushed by miners and equipped with wheelsets—that is, wheel assemblies in which the wheels are fixed to the ends of the axle, with the entire assembly rotating together.

Writings on the origin of the wheel are voluminous, but they usually present a catalog of archaeological finds and steer clear of firm conclusions about attendant circumstances and specific places of origin. Stuart Piggott, the most renowned archaeologist to take up the question, expressed dismay over what he felt was the near simultaneity of the earliest evidence for wheel use in Mesopotamia and in northwestern Europe:

What can be said is that any time interval between the present evidence for the first wheeled transport in north-west Europe and Mesopotamia must be very short, and we can only fall back on inherent historical and technological probability in assuming a more likely initial invention as a part of the complex innovations of this period [ca. 3000 B.C.E.] in the Near East rather than in the simpler complex of Neolithic Europe.[1]

Carbon-14 analysis brings this "inherent historical and technological probability" into question. Since 1950, archaeologists have used the carbon-14 found in organic material from their excavations to estimate the age of the sites. Carbon-14 is a naturally occurring radioactive isotope that decays at a predictable rate, and the decay begins upon the death of the living matter that absorbed the carbon from the atmosphere. The extent of the decay provides a measure of the number of years that have elapsed between the onset of the decay and 1950, the base-line year used in these studies. Subsequent studies of specific datable pieces of wood, however, found that the level of carbon-14 in the atmosphere, assumed in 1950 to be constant, has steadily diminished over time, though without perfect regularity. Through measuring the decay of carbon-14 in the rings of a long-lived tree, each ring being composed of dead matter, it became possible to "correct" or "calibrate" carbon-14 dates and bring them into closer agreement with calendrical dates. That is, a carbon-14 date of, say, 5050 B.P. (before present), the equivalent of 3100 B.C.E., may actually represent a calendrical year around 5750 B.P. (3800 B.C.E.) once it is corrected by tree-ring calibration.

Piggott was well aware of the uncertainties in carbon-14 dating and of its calibration by tree-ring analysis. "In this book," he wrote, "the practice of using uncalibrated radiocarbon dates has been followed. . . . Where calibrated dates, giving a higher antiquity, have to be considered is in contexts where comparison is needed between radiocarbon and historical chronologies."[2] The practical effect of his disinclination to accept calibrated carbon-14 dates was to favor dates for the earliest surviving wooden wheels, seventeen fragments of which come from sites in northern Switzerland, southern Germany, and Slovenia, that are several centuries later than the calibrated dates used by other archaeologists. His reservation about tree-ring calibration caused him to see little or no time interval between the earliest evidence from Mesopotamia, which took the form of written ideograms, and the uncalibrated radiocarbon dates of the wheels found in "north-west" Europe.

The calibrated dates that Piggott did not use indicate that some Europeans were using wheels well before their first appearance in Mesopotamia. But this conclusion flew in the face of the Mesopotamian-origin theory that he felt was correct. His insistence on the "inherent historical and technological probability" of Mesopotamia being the birthplace of the wheel has gained wide acceptance, even though no physical evidence confirming the theory has ever been found. After all, if, as is generally assumed, civilization took a giant step forward with the invention of the wheel, then it seems only logical that this step should coincide with other innovations being made by the world's first urbanized society in the valley of the Tigris and Euphrates Rivers. To think otherwise, to think that the wheel may have begun as a local solution to a particular transportation problem, seems like an anticlimax. Yet I believe that that is exactly what happened in the copper mines of the Carpathian Mountains. Let us take a look at the evidence.

I begin with another potter putting wheels on an animal figurine, but this one is molding his clay almost four thousand years before the potter in southern Mexico. The location is in the Carpathian Mountains of western Ukraine, and the date, based on calibrated carbon-14 analysis, is between 3950 and 3650 B.C.E., about a thousand years before the Egyptians built their first pyramids. The term "Old Europe" has come into use for the cultures of southeastern Europe in the fourth millennium B.C.E. The animal is a bull instead of a dog, but the wheels are the same simple disks used by the Olmec toy maker (figure 26).

A tally of early evidence for wheels encompassing thirty-five archaeological sites from Denmark to Pakistan points to this zebra-striped figurine as the earliest known example of an object on wheels. Does this mean that the person who crafted it had seen an actual vehicle? As with

FIGURE 26 A wheeled bull figurine from the Cucuteni-Tripolye culture of western Ukraine. (Museum of Historical Cultural Heritage PLATAR, Kiev)

the pre-Columbian toy, there is no way of determining whether tiny wheels on a clay figurine indicate anything one way or another about a society's use of big wheels, particularly since big wheels made of wood rarely survive archaeologically. So let me ask instead a different question: Did the region of the Carpathian Mountains witness any change in its transport needs around 4000 B.C.E.?

The answer to this question is yes. The Copper Age, an era in human history that preceded the Bronze and Iron Ages, had begun around 5500 B.C.E. in Serbia, in southeastern Europe. Over the next 1500 years, through the Early and Middle Copper Ages, artisans fashioned many types of copper objects. Gold objects also began to appear, perhaps refined from the same ore, since copper and gold melt at about the same temperature. By the Late Copper Age, however, the number of copper objects declined, probably because easily collected surface deposits of malachite and other copper ores were exhausted. Mining afforded an alternative to collecting surface ores, but mining required driving tunnels into the sides of mountains, possibly in pursuit of veins of ore that were less rich than what the metalworkers were used to. It is in this context, the Carpathian Mountains during the Late Copper Age, that evidence of wheeled vehicles first appears in quantity.

Although less familiar to travelers than the Alps, the Carpathians constitute one of the great mountain ranges of Europe. They extend from Ukraine in the east to the Czech Republic in the west and cover parts of Poland, Slovakia, Hungary, Serbia, and Romania. A map of where copper was mined in Europe during the Copper Age shows some production in Britain and Spain; more widespread production in the Balkans, especially in Serbia and Bulgaria; and a heavy concentration farther to the north following the arc of the Carpathians (figure 27). Elsewhere, copper was also mined in eastern Turkey, on Cyprus, in northern Iran, and in the Caucasus Mountains.

Once the most productive surface ores were exhausted and mines began to be bored into mountainsides, miners in all these areas faced a transportation problem: What was the best way to get their ore out of the mine and move it to a fire that would smelt the copper out of the ore?

Copper ore is dense, weighing about 140 pounds per cubic foot, but most of it is waste. A cubic foot yields only 1 to 3 pounds of copper. Thus the transportation problem involved carrying an enormous weight of chipped stone, most of it waste, along a stone tunnel to the mine's entrance. The collection of surface ore probably had involved carrying baskets across rough ground. But the corridor of a mine was hewn out of stone by the miners and could be made flat and straight. A rough surface of dirt, stone, and grass would still be encountered outside the mine, particularly since the fires used for smelting might move from place to place as the smelters cut down all the nearby trees. (One Irish mine of the Copper Age, which is estimated to have produced almost 36,000 tons of ore, must have caused whole forests to be felled.)

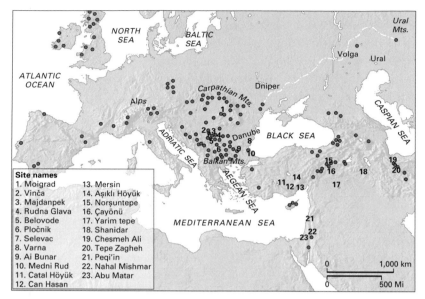

FIGURE 27 Sources of copper during the Copper and Bronze Ages. (Redrawn from Ernst Pernicka and David W. Anthony, "The Invention of Copper Metallurgy and the Copper Age of Old Europe," in *The Lost World of Old Europe: The Danube Valley 5000–3500 B.C.*, ed. by David W. Anthony, with Jennifer Y. Chi [Princeton, N.J.: Princeton University Press, 2010], 164)

The earliest known miners worked in the Old Stone Age, many thousands of years before the discovery of copper. They dug for minerals and rocks like flint and obsidian, which were good for fashioning tools and weapons. But their mines yielded far less waste because every basket load that the miners excavated was valuable. By comparison, copper mining meant lots of weight and very little pay-off per basket of ore. Yet as surface collection gave way to the greater difficulties of underground mining, and thereby reduced the quantities of copper that could be produced, the value of the copper probably rose.

A SQUARE PEG IN A ROUND WHEEL

Of course, like that of manure or hay, the weight of the copper ore could be distributed—and undoubtedly was in most places—over numerous human porters. But probably not over pack animals. Oxen were the only available beasts of burden in Copper Age Europe, and oxen could not readily fit into mine tunnels, especially if the entrance was a vertical shaft. So for the miners working inside, lugging their trays or baskets of ore along a corridor may have seemed like a poor expenditure of muscle power that might be better used attacking the mine face with an antler or a stone pick (figure 28).

This, I am proposing, is what some unknown miner in a Carpathian mine was thinking about one day when he fastened onto the idea that it would be easier to shovel the ore into a bigger basket mounted

FIGURE 28 Miners using baskets to carry ore in an ancient Greek mine. (Wikipedia Commons)

on wheels and then roll the basket back to the mine's entrance. This hypothesis, even though it applies to only one copper-mining region, links an increasingly burdensome task in the mining of copper ore to the earliest, most detailed, and most abundant archaeological evidence for the first appearance of wheeled transport.

Was copper mining the mother of the invention of the wheel? What is the evidence? Shortly after the crafting of the wheeled bull figurine in Ukraine, and in the same general time period as the production of the full-size wheels whose fragmentary remains have been found in northern Switzerland, southern Germany, and Slovenia, potters working at sites on the southern flank of the Carpathian Mountains produced many clay models of four-wheeled cars. To date, at least 150 examples have been found. This, then, appears to be the original home of the wheel.

The earliest wheeled vehicles probably looked roughly like that shown in figure 29 . . . except for the big loop handle. For like the other clay models, this is a drinking mug a couple of inches wide and not an actual mine-car. Mária Bondár, a Hungarian archaeologist specializing in the Copper Age and in early wheeled transport, has devoted a book and numerous articles to a painstaking study of these clay vehicle models (with and without mug handles) and how they relate to the examples of actual wheels that have survived from the Copper Age:

> The Late Copper Age wagon models have a rectangular wagon box with trapezoidal sides (the top being longer than the bottom) and an open top. The differences in their ornamentation and their rim form suggest that genuine wagons too were made from different materials using diverse techniques. . . . The many wooden finds discovered in [recent years in Switzerland, Germany, and Slovenia] have furnished incontestable proof that the various components such as the wagon box, the axles, and the wheels had been made from wood. While the *axle and the wheels rotating with the axle* had been made from planks, a much wider range of materials were probably employed for the wagon box such as wood (planks of varying length), wickerwork reinforced by rods, or a combination of the two.[3]

FIGURE 29 A clay mug in the form of a four-wheeled vehicle from the Boleráz culture. (László Gucsi, "Copper Age wagon model from Grave 177 of the Budakalász cemetery," in Mária Bondár, *Prehistoric Wagon Models in the Carpathian Basin [3500–1500 BC]*, Series Minor 32 [Budapest: Archaeolingua, 2012], fig. 5. Used with permission)

Although Bondár does not suggest a connection between the clay models she has studied and copper mining, she does observe that none of those currently known come from the flat grasslands of the Great Hungarian Plain, which constitutes much of Hungary and extends into neighboring countries and where one might expect wagons to have been particularly

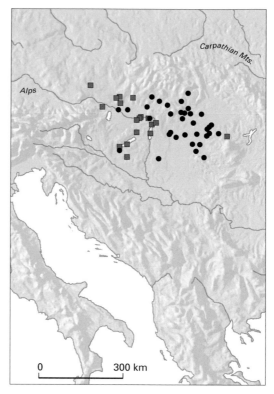

FIGURE 30 Archaeological sites from the Boleráz culture where clay models of wheeled vehicles have been found. Squares = Copper Age; circles = Bronze Age. (Modified, with permission, from Mária Bondár, *Prehistoric Wagon Models in the Carpathian Basin [3500–1500 BC]*, Series Minor 32 [Budapest: Archaeolingua, 2012], figs. 19 and 33)

useful (figure 30). Crucially, with respect to the competition between wheel concepts discussed in chapter 1, she also notes that wheelsets were used instead of wheels rotating independently on the ends of an axle.

The technical name that archaeologists assign to the society in which the clay models first appeared is the Boleráz phase of the Baden Culture, a pattern of living that ranged eastward from Austria and extended into Hungary on its eastern frontier. Until recently, archaeologists who had noted certain similarities between the Boleráz and other early cultures in Greece, Romania, and Bulgaria had theorized that there probably had been a northwestward diffusion into Hungary of techniques and practices from southeastern Europe, or perhaps from even the Near East. This theory fit nicely with Piggott's belief that the wheel was invented in Mesopotamia. Then, starting in 2002, comprehensive comparisons of calibrated carbon-14 dates from the earliest Boleráz sites with dates from similar sites farther south began to show that the Boleráz culture took form no later than 3600 B.C.E., and thus was older than the cultures that supposedly had influenced it.[4] Therefore, it now seems that any Late Copper Age diffusion of cultural practices in eastern Europe, including the use of wheels, went from north to south, from Hungary to Greece, rather than vice versa. Piggott's favored northward diffusion from Mesopotamia seems to be ruled out.

The Lengyel Culture, also centered in Austria, immediately preceded the Boleráz phase of the Baden Culture, but the two did not differ very much. Both peoples lived in villages; both grew wheat and other grains; both kept cattle, pigs, and sheep. But the Lengyel people used much less copper than the Boleráz, mostly for beads and hammered strips—maybe because it was becoming harder to mine the ore. Whatever the case, wheeled mine-cars, which would have increased production, were probably invented by the Lengyel predecessors of the Boleráz folk.

Ascribing the invention of the wheel to a late Lengyel date closer to 4000 than to 3600 B.C.E., when the Boleráz culture took form, seems logical even if there is little in the way of concrete evidence to support it. The abundance of clay models proves that wheeled vehicles played not just a central role in the Boleráz culture, but an iconic one as well. We may not know what kind of rituals the vessels were used for—conceivably, they were just souvenirs—but obviously people thought they were important, and they would not have made so many models if the wheel had been invented just a few years earlier. It takes time for objects to acquire enough symbolic significance to become cultural icons. Witness, for example, the century-plus that elapsed between Richard Trevithick's tentative invention of a steam car in 1803 and the emergence of the automobile as an icon of modern society in the World War I era. A Lengyel miner may have invented the wheel, but the Boleráz people expanded its role in the economy and adopted its image as a cultural icon.

Nevertheless, although many scholars in addition to Bondár have written a little or a lot about the Boleráz clay models, no one has connected them with mining. Indeed, they have paid greater attention to two other traces of early wheeled transport. One is a design scratched around the rim of a pot, carbon-dated to between 3631 and 3380 B.C.E., found at Bronocice in southern Poland, just north of the Carpathian Mountains. The design shows a three-pronged projection from one end that seems to represent a wagon tongue and a yoke for harnessing oxen (figure 31). This suggestion of a means of harnessing animals is absent from all the Boleráz clay models. Some models do have two animal heads—probably cattle but lacking the horns—projecting as knobs from one end, but none gives any indication of how, if they were used for pulling, these animals may have been hitched to the vehicle.

The other vestige of the use of wheels on vehicles, which is contemporary with the period of the earliest Boleráz clay models, is a sixty-five-foot-long pair of ruts underlying a burial barrow at Flintbek, a site in Germany far north of the Carpathians near the border with Denmark. The ruts, which date to between 3460 and 3385 B.C.E., run in a straight line and are three and a half feet apart. There is no way of telling whether the ruts were made by two wheels or four, or even by the flat runners of a sledge, but the space between them is rather narrow considering the width of a serviceable farm vehicle.

Returning to the clay models, Bondár's description of them as having high sides that often lean outward beyond the length of the axles is incomplete. Although many of them have only vestiges of the four projections underneath that originally had holes through which the axles passed, like the paws of the Olmec potter's toy dog, the most complete models have fairly small wheels. Their diameters are only one-third to one-half as great as the models are high, and the length of the models

FIGURE 31 A representation of a four-wheeled vehicle scratched into a pot found at Bronocice, Poland, with an indication of a wagon tongue and a yoke for oxen. (Modified from Stuart Piggott, *The Earliest Wheeled Transport: From the Atlantic Coast to the Caspian Sea* [Ithaca, N.Y.: Cornell University Press, 1983], figs. 10 and 11)

is roughly two and a half times the diameter of the wheels, with the width being slightly shorter. These proportions suggest that the size of the wheels might be used to determine the size and shape of the vehicles that the models represent. For example, if the models were patterned on a real vehicle with wheels three feet in diameter, which is a fairly standard diameter for later wagon wheels, then the sides would have been an impractical six to nine feet high, with no indication of an entrance. How would one get in? How would one get anything out? So it is likely that the wheels were smaller. But how small?

The full-size wooden wheels of a similar age whose fragments have been preserved in northern Switzerland, southern Germany, and Slovenia have square holes in the center, showing that, like the wheels on the clay models, they were fixed to the ends of axles rather than freely rotating. The largest of a group of four from Germany has a diameter of twenty-three inches, and an even earlier specimen from Slovenia, also with a square hole in the middle, has a diameter of twenty-eight inches (figure 32). None of these examples of actual wheels indicate an association with mining, though mining did take place in the Alps as well as the Carpathians in the Copper Age. Nor do they fit the proportions of

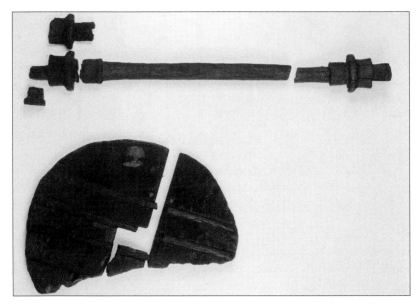

FIGURE 32 An early wooden wheel and axle from Slovenia. The square hole in the middle confirms that the wheels were fixed to the axle and did not rotate independently. (Collection of City Museum of Ljubljana, Slovenia)

the Boleráz models. That is, if a real vehicle had a 1:2.5 ratio between wheel diameter and height, a twenty-eight-inch wheel would produce a wagon box almost six feet high. So if wheelset vehicles did originate as mine-cars, as I am proposing, the concept must have spread quickly into flat country outside the mining regions.

Since the surviving wheels are bigger than what the Boleráz models would suggest, we can only guess at the dimensions of the mine-cars that the clay mugs represent. Given the proportions of the models, a twelve-inch wheel would indicate a mine-car that was two to three feet deep, with a wheelbase of two and a half by two feet. As for the material used to construct the body of the car, Bondár's suggestion of wickerwork with a solid plank floor fits best with the stylized patterns of lines on all four sides of many models (figure 33). This probably reflects the well-attested

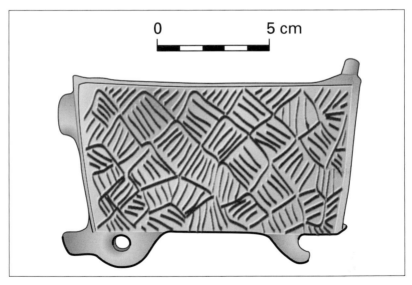

FIGURE 33 The wickerwork pattern on a clay model of a mine-car from the Boleráz culture. (Redrawn, with permission, from "Copper Age wagon model from Moha," in Mária Bondár, *Prehistoric Wagon Models in the Carpathian Basin 3500–1500 BC]*, Series Minor 32 [Budapest: Archaeolingua, 2012], fig. 12)

use of baskets in ancient mines (see figure 28). However, extraordinary as it may seem, coach builders in Hungary, though not anywhere else, were still—or again—using reinforced wicker for the sides of their vehicles in the fifteenth century, almost five thousand years later (see figure 78).

Although the diameters of the earliest surviving wheelsets from Germany, Switzerland, and Slovenia are large enough to accommodate modern notions of what a traditional farmers' wagon should look like, the overall size and shape of a full-size Carpathian mine-car was probably comparable, in today's terms, to a wheeled laundry cart or household refuse cart (figure 34), albeit with solid wooden wheels instead of casters (chapter 11).

FIGURE 34 A modern refuse cart. (Photograph in author's collection)

Miners did not need large vehicles. Filled with heavy ore, they would have been too hard to move. So depictions of mine interiors, which began to become available many millennia later, such as in Georgius Agricola's *De re metallica*, a sixteenth-century treatise on mining in central Europe, typically show hand-pushed mine-cars that in size and shape are quite similar to what the clay models indicate a Boleráz mine-car would have looked like if it had wheels that were one foot in diameter (figure 35). Nor has mine-car design changed much since the sixteenth century, as we can see in a photograph taken around 1900 (figure 36). The cars now run on tracks, but they still have wheelsets with axles that go through holes in brackets attached to their undersides.

Bringing up images of mine-cars from the sixteenth and nineteenth centuries to support an argument for the invention of the wheel more than five thousand years earlier may seem far-fetched. After all, the surviving wooden wheels that date to the same period as the Boleráz clay models were not found in the Carpathian Mountains and have dimensions that are more appropriate to farm wagons than to mine-cars. Moreover, the models themselves have no explicit association with mining.

Yet the steering problem discussed in chapter 1 keeps my argument on track. All the clay models have four wheels. In keeping with Bondár's judgment, and taking note of the knobs with axle holes that protrude from the models' undersides, it is apparent that the vehicles they imitate were equipped with wheelsets rather than with wheels that rotated independently of one another. The square holes in the center of all the wheels that survive from the period confirm this design, and no example of wheels rotating independently on the ends of their axles can be dated archaeologically to this early period, either in a model or in an actual wheel.

This means that none of the vehicles that the clay models were patterned on could be steered. A four-wheeled wagon rolling on two wheelsets would have been as unsteerable as a railroad car, and no ancient example of such a vehicle outside the context of mining has ever come to light. Thus the models reflect a pattern of use in which steering was not needed, but four wheels were. This clearly points to

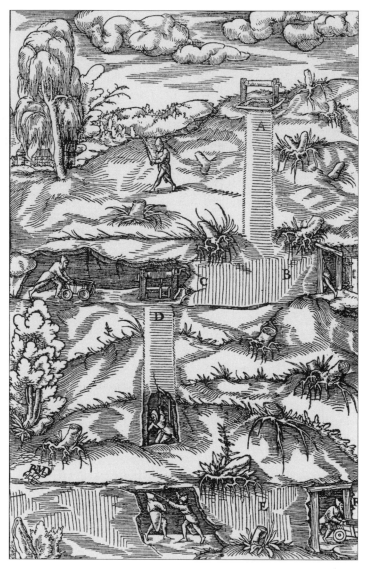

FIGURE 35 The representation of a mine in Georgius Agricola, *De re metallica* (1556), showing mine-cars at middle left and lower right. (Copyright © World History Archive / Alamy)

FIGURE 36 Hand-pushed mine-cars on rails, ca. 1900.
(Copyright © The Protected Art Archive / Alamy)

a mining environment, both because the miners could have fashioned straight tunnels with flat stone floors along which to push their ore-cars and because the weight of the ore would have made a two-wheeled cart almost impossible to balance. The survival down to modern times of small, hand-pushed mine-cars running on wheelsets confirms the utility of such vehicles in—and only in—mines.

The spread of the wheel concept outside the mines probably occurred fairly quickly in the form of two-wheeled oxcarts, and the full-size wheels that are contemporary with the Boleráz clay models most likely belonged to such carts. They did not take on iconic significance, however, probably because they were not terribly efficient. It is difficult to turn a full-size, two-wheeled oxcart running on one wheelset. It has to be partially dragged around curves because the outer wheel cannot make more

FIGURE 37 An oxcart in central Turkey, equipped with a wheelset in which
the axle rotates in grooves cut underneath the frame of the cart.
(Photograph in author's collection)

revolutions than the inner wheel. But the survival of oxcarts with single
wheelsets in such places as Portugal, Spain, and Turkey into the twentieth century proves that this was not an insuperable problem (figure 37).
In addition, the axle need not go through holes in projections underneath the bed of the cart. Instead, a thick axle can rotate in grooves,
which causes the loud squeaking that nineteenth-century travel writers
frequently mentioned. The only advantage the oxcarts with wheelsets
had over the steerable oxcarts with independently rotating wheels that
almost completely superseded them as the Copper Age gave way to the
Bronze Age was simplicity of construction. A square hole in a round
wheel sufficed to fix solid wooden wheels to the ends of an axle, but crafting wheels to rotate independently was substantially more complicated.

four

Home on the Range

Judging from calibrated carbon-14 dates, several centuries elapsed between the invention of the wheelset and that of wheels that rotate independently at the ends of an axle. Southeastern Europe is the likely locale for the latter invention, presumably achieved by people who had seen or heard about wheelset vehicles in use in the Carpathian Mountains farther to the north. Archaeological and linguistic evidence for the wide-scale adoption of the independently rotating wheel, however, points to the plain on the northern shore of the Black Sea in present-day southern Ukraine and southern Russia. This plain and its eastward extension into the valley of the Volga River is dotted with ancient burial mounds, and 250 of these mounds, dating between 3000 and 2000 B.C.E., contain the remains of four-wheeled wagons (figure 38).

This chapter proposes that these wagons were buried with their owners because they had served during their life—and possibly were meant to do so as well in their afterlife—as mobile homes, the ancestors of these people having adopted a nomadic style of life after being uprooted as farmers by severe climatic and hydrographic events. In other words, if the needs of miners proved to be the mother of invention for the

FIGURE 38 A four-wheeled wagon that was interred in a burial mound north of the
Black Sea between 2300 and 2200 B.C.E. The protruding naves in the
center of the wheels prove that the wheels rotated freely on their axles.
(Photograph in author's collection)

wheelset, a mobile-home lifestyle seems to have played a similar role in popularizing the use of independently rotating wheels.

It is not surprising that rotating wheels appeared later than wheelsets. They are more complicated to build because instead of just fixing a wheel so that it does not move, which is easily done by making the end of the axle square and inserting it into a square hole in the center of the wheel, a rotating wheel must turn freely. If the round hole that the axle goes through in the center of the wheel is roomy enough for the wheel to turn without much friction, however, a narrow wheel will inevitably tilt to the side a bit and wobble when it rolls, eventually wearing through the axle. A thicker wheel will not wobble but may be too heavy for practical use.

Take oak, for example. Many of the earliest wheels were made of solid oak. A cubic foot of air-dried white oak weighs 47 pounds. Thus a solid wheel that is two feet in diameter and one inch thick weighs 11 pounds. This wheel might serve perfectly well fixed to the end of the axle in a wheelset, but it is not thick enough to keep from wobbling if it is supposed to rotate on its axle. Increasing the depth to three inches brings the weight of the wheel to 35 pounds, which is about 14 pounds heavier than today's average car tire. But that is still not enough to prevent it from wobbling. An early rotating wheel had to be at least six inches thick to turn smoothly. For a solid oak wheel six inches thick, this brings the weight up to 70 pounds, as much as three and a half modern car tires, or 280 pounds for a four-wheeled wagon. Add a few hundred more pounds for the wagon's body, axles, beam, and yoke, and the vehicle is altogether too heavy to be of much use.

To hold a wheel to an acceptable weight, the craftsman who fashioned it made the center much thicker than the edge. The rudimentary way of doing this was to start with a plank thirty inches wide and six inches thick and then chip away most of the wood until the outer rim of the wheel was two to three inches thick, but there was a sort of tubular sleeve in the center for the axle to pass through (see figures 38 and 39). This protruding sleeve is called a nave, and whether it was made from the same piece of wood as the rest of the wheel or, in a later development, was fashioned as a separate piece and inserted into a large hole in the wheel's center, it is as much the hallmark of early independently rotating–wheel systems as a square hole in the center of a wheel is the hallmark of an early wheelset. Naves usually show up clearly in miniature models from different early cultures that used independently rotating wheels.

All the wagons found in the burial mounds north of the Black Sea have wheels with protruding naves, which makes them both visibly and conceptually different from the wheels found in Switzerland, Germany, and Slovenia dating to the earlier Borenáz phase of the Baden Culture (see figure 32). The diameters of the wheels are also bigger, and the wagon frames are more rectangular than those of the Borenáz clay models. (The physical remains of early wheelset vehicles in Switzerland, Germany, and Slovenia are too fragmentary to calculate the size of the vehicles to which they belonged or even the number of wheels they had.) The Black Sea wagons had lower sides than what I have deduced for the Carpathian mine-cars, and the sides did not tilt outward. Most important, however, they were not pushed by hand. Motive power came

FIGURE 39 A clay model of a cart from the Indus Valley civilization, second millennium B.C.E., whose wheel has a nave in the center. (National Museum of Pakistan, Karachi. Universal Images Group / Art Resource, NY)

from pairs of oxen harnessed to the wagons by a yoke, a wooden cross-piece tied to a beam projecting from the center of a wagon's front end. The animals were sometimes buried with the wagons and their owners. Putting these details together, you get a Black Sea vehicle that looks like a smallish farm wagon rather than a hand-pushed mine-car.

Even with independently rotating wheels, however, the steering problem inherent in wheelset vehicles was not solved. The wagons that trundled at about two miles per hour across southern Ukraine and southern Russia had four wheels. Two-wheeled carts, which can steer more easily because the outer wheel is free to make more revolutions than the inner wheel, seem not to have been used in the region until around 2000 B.C.E., a full millennium after the four-wheeled wagons appeared. The new wheel design allowed the four-wheeled wagons to round curves that would have been impossible for a four-wheeled mine-car equipped with two wheelsets to maneuver, but the curves they could manage were still very gradual. If the wagons were basically mobile homes, however, they may not have had to move very far on a given day, and the Black Sea plain was generally quite flat.

The solution to the steering problem that would eventually be adopted was to attach the middle of the front axle to the bottom of the wagon with a thick metal pin. This allowed the front axle to pivot and the front wheels to change direction until the edge of one or the other wheel rubbed against the body of the vehicle. However, there is no sign of a pivoting front axle in any of the Black Sea burials. This may be because a thick piece of metal would have been needed for the pin, and the bronze that came into use with the waning of the Copper Age may have been too scarce or too weak to do the job. In any case, people wending their way slowly across a trackless countryside may have been happy to go straight ahead most of the time.

The person who has explored the evidence for wheel use in the Black Sea plain most intensely is the anthropologist David W. Anthony. In chapter 2, I took issue with Anthony's observation that wagons must have been especially useful for carrying "harvested grain crops, hay crops, manure for fertilizer, firewood, building lumber, clay for pottery making, hides and leather, and people."[1] But this objection in no way diminishes the importance of his work. It simply raises a question that he chooses not to address: How likely is it that a prehistoric Ukrainian farmer who had used a wagon for much of his life to carry hay, manure, and firewood would have wanted it to be buried with him?

There is no disputing Anthony's common-sense assertion that wheeled vehicles are useful to farmers, but something different is needed to explain their inclusion in burial mounds in Ukraine and Russia. The question that must be asked is why these wagons were buried, and the answer I am proposing is that they were interred with the bodies of the people who had lived in them, whether or not they had also used them for farming chores. The reasoning behind this answer involves a sequence of hydrographic and climatic changes that had a unique impact on the

Black Sea plain and forced at least some of the people who lived there to adopt a nomadic lifestyle.

The glaciers of the final Ice Age melted steadily after around 9000 B.C.E., and the balmy temperatures that caused the melting stimulated the development of village and farming life in southeastern Europe and elsewhere. The period is sometimes referred to as the Climatic Optimum. Then four thousand years or so after the glaciers began to shrink, somewhere around 5000 B.C.E., the weather changed. Warm and moist gave way to cold and dry, with a negative impact on fledgling agricultural communities in many parts of the world. In some areas, such as the Sahara and the Arabian Peninsula, immense deserts slowly overtook what had been grasslands, and rivers and lakes went dry. Another bout of cold and dryness set in after 4000 B.C.E., and the Sahara reached its current level of extreme dryness around 2500 B.C.E.

In Bulgaria and Romania, the cultures of Old Europe fell into decline around 3900 B.C.E., when wheelset vehicles were first coming into use farther to the north in the Carpathian Mountains and the Alps. Most villages disappeared, and the Balkan copper mines, which had once been very productive, ceased operating. Archaeologists do not agree on why this previously flourishing region, which bordered the northwestern edge of the Black Sea plain, suffered such profound collapse, but a similar decline affected farming villages in northern China in the same period.

As for the Black Sea plain itself, as well as adjoining parts of Central Asia to the east of the Volga River, forests disappeared and grasslands took their place after about 3300 B.C.E. The steppes of historical times came into being in this period. But this was not the only change in the Black Sea area. Sometime around 5000 B.C.E., the water melting from the world's glaciers had raised the global sea level enough for the Mediterranean Sea to cut through a land barrier that had long separated it from the Black Sea. The opening of a spillway next to the Bosporus, a strait that today provides a channel to the Black Sea between the European and Asian sides of Istanbul, caused an immense flood that inundated a huge area on the Black Sea coasts of what are now Romania, Ukraine, and Russia. Before this flood, the Black Sea was a freshwater lake with a surface some

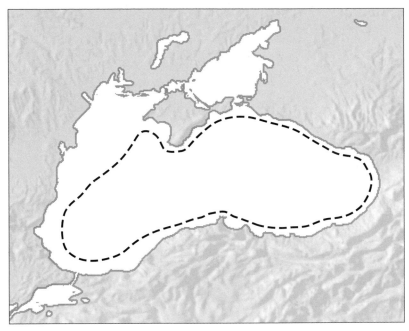

FIGURE 40 The area of the Black Sea today (*white*)
and in 5500 B.C.E. (*within dashed line*).

hundred feet lower than it is today (figure 40). When the Mediterranean waters broke through, however, an unimaginable torrent at the western end of the Black Sea filled it with salt water. Farming villages, which were probably few and far between, sank beneath the waves across tens of thousands of square miles. Most of the evidence of this human calamity disappeared, of course, but underwater archaeology has uncovered the remains of one fifth-millennium B.C.E. village at a depth of eighteen feet. The Black Sea's expansion stabilized between 4500 and 4300 B.C.E. and then resumed at a slower pace between 4200 and 3600 B.C.E.

The 250 burial mounds that include wagons are located in the region north of the new coastline and extend eastward to the north of the Caucasus Mountains and up the valley of the Volga River (figure 41). Why

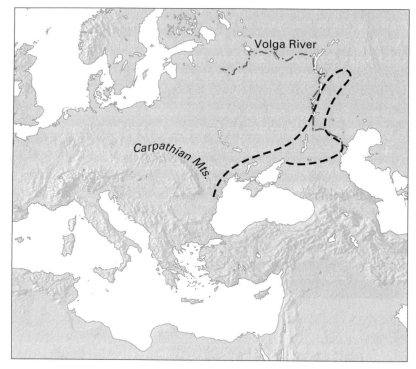

FIGURE 41 The region where 250 burial mounds with interred wagons, dating to before 2000 B.C.E., have been found (*within dashed line*).

they were buried will never be known with certainty, but it is reasonable to assume that they reflect a ritual of some sort, indicate the social status of the deceased, or were intended for the deceased to use in the afterlife. The wagon, with or without its draft animals, is often the sole item distinguishing one burial from another. Being large and costly, the wagon probably was not built for just a funeral procession. More likely, it reflects some aspect of the earthly lifestyle of the deceased.

The Black Sea flood of 5000 B.C.E. probably had been totally forgotten, or perhaps turned into myth or legend, by the time the wagon burials began two thousand years later. But the dislocation caused by the rising waters had forced many villagers to leave their land and villages forever. And the flood had never receded. This may have implanted a preoccupation with mobility in the minds of the farming people of the region, a concern that someday the sea might again overwhelm them. And it may have seemed to be recurring after 4200 B.C.E.

The buried wagons were practical vehicles: unornamented, heavy, and slow. And they could make only gradual turns. There was nothing flashy about them or about the mobility they afforded. Even if they had not been interred with their owners, they would seem better suited for personal transport than for farm chores. Two-wheeled carts, which people in the Black Sea region seem never to have thought of until a thousand years after the wagon burials began—that is, until around 2000 B.C.E.—would have been far more practical than four-wheeled wagons for hauling hay and lumber. A cart with independently rotating wheels not only is maneuverable, but, using the same pair of oxen, can actually carry a heavier load than a four-wheeled wagon, since two wheels turning on a single axle generate only half the friction produced by four wheels turning on two axles. But people could live inside a four-wheeled wagon, with a cloth or leather top stretched over light overarching poles, and they could not so easily inhabit a cart.

Before discussing how vehicular nomadism developed in later times, however, two matters relating to the buried wagons have to be considered: the words employed by the people who used and buried them, and the way those people changed their method of crafting wheels between 3000 and 2000 B.C.E.

Language is at the center of Anthony's compelling argument for the Black Sea plain being the original home of the peoples who spoke what linguists call Proto-Indo-European (PIE). Linguists have reconstructed this language hypothetically from the vocabulary and grammar of the scores of later Indo-European tongues—including English, Greek, Hindi, Latin, Persian, and Russian—that ultimately derive from PIE and are now spoken from Ireland to Bengal and throughout the Western Hemisphere. Without entering into the vigorous debate among rival theories concerning where PIE was originally spoken, I shall attend to only what Anthony, in agreement with every other specialist on the Indo-European language family, has to say about wheels: that all the languages in the family share the roots for certain words relating to vehicles with independently rotating wheels. In English, these roots produced the words "axle," "cart," "nave," "road," "wagon," "wheel," and "yoke."

Note, however, that there is no common word in any Indo-European language for wheels that are fixed to the ends of an axle. The English word "wheelset" seems either to have been coined at the beginning of the railroad era or to derive from a technical mining term that has left no earlier dictionary trace. All PIE wheels rotate freely on the ends of an axle. The word "nave" clearly signifies this, since wheelsets do not require naves. Moreover, the most common PIE root for words meaning "wheel," the root from which the English word "cycle" derives, comes from a PIE verb stem meaning "to turn or revolve." Wheels revolve; axles do not. Hence it seems likely that the people who spoke PIE either were unaware of the earlier fixed-wheel alternative to the fully rotating wheel or considered it an inferior technology. That is why I have proposed that the independently rotating wheel was probably invented somewhere in southeastern Europe intermediate between the Carpathians and the Black Sea. Cultural influences moving from northwest to southeast show that people in, say, Bulgaria or Macedonia had contact with the Carpathian region, which is where the Bordenáz people used wheelset-type vehicles and presumably devised technical terms in their own non–Indo-European language. Regardless of where the idea of wheels rotating on the ends of axles originated, however, when the PIE speakers of the Black Sea plain adopted the technology, they came up with a distinct set of words to describe its main features.

For Anthony, the words relating to wheels that are shared by the languages descending from PIE support his idea that the Black Sea plain was the PIE homeland. But the absence of words can be as suggestive as their presence. Not only are there no words pertaining to the earlier use of axles with fixed wheels, but the Indo-European languages do not share words referring to changes in wheel construction that culminated in the invention of wheels with spokes shortly before 2000 B.C.E.

Spokes, which were initially made of wood, connect the center of the wheel with the rim. One end of each spoke slots into a hole in the hub, which functions both as a nave to keep the wheel from wobbling and as the central support of the entire wheel. The other end of the spoke slots into the rim, which was originally constructed of several curved pieces of wood pegged together end-to-end to make a circle (figure 42). Thus in addition to "spoke" and "hub," English has a third key term for the rim

FIGURE 42 The construction of wooden spoked wheel. Two spokes connect each felloe to the hub. (© Can Stock Photo Inc. / piai)

pieces. Each one is called a "felloe" (usually pronounced "felly"), but the word fell out of common use after rims began to be made of solid pieces of metal.

That the different Indo-European tongues have entirely distinct words for these new wheel parts implies that the dispersal of the Indo-European peoples from a place of origin, where they utilized a common type of solid wheel with a nave in the center, began after independently rotating wheels came into common use, but before spoked wheels were invented. This implies that before 2000 B.C.E., the speakers of PIE had begun to break into groups and migrate in different directions and that the wagon builders in each migrating group implemented the hub-spoke-felloe concept in different ways, coining different words in the process. This fits with the large and diverse array of archaeological evidence for

FIGURE 43 A Viking wagon that was buried in Norway, ninth century. (Photograph by Annie Dalbéra. UIO Museum of Cultural History, The Viking Ships Museum, Oslo)

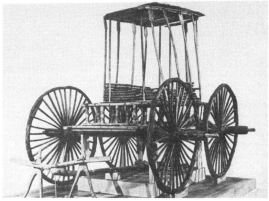

FIGURE 44 A wagon that was buried at Pazaryk, Siberia, fifth century (*top*), and its reconstruction (*bottom*). (Hermitage Museum, St. Petersburg, Russia)

the construction of spoked wheels. The wheels on a wagon buried in Norway in the ninth century C.E., for example, have six thick felloes and twelve short spokes—two per felloe—while a wagon buried in Siberia, dated four centuries earlier, has wheels with thirty-three thirty-inch-long spokes and only two thin felloes made of bent wood (figures 43 and 44).

Why were spoked wheels invented? Since solid wheels were still widely used on oxcarts in different parts of the world at the turn of the twentieth century, the spoke cannot be considered essential to wheeled transport. However, it was essential to the development of fast, lightweight war chariots after 2000 B.C.E. So did people invent spokes with chariots as their goal? Or did they invent spokes for other reasons and then discover that a cart with spoked wheels could do things that had never been thought of before?

We know from the earliest surviving specimens that both fixed wheels and independently rotating wheels were chipped out of one to three thick planks of wood. (Contrary to a common misconception, cross-sections of tree trunks have seldom been used as wheels.) When three planks were used, they were fastened together by mortise and tenon—that is, tab A in slot B—and glue, and sometimes were reinforced by thin strips of wood going crosswise instead of radially (see figure 32). The word for such a crosspiece is "batten." The planks that the wheelwright started with had to be wide and thick. For a wheel thirty inches in diameter, he needed a tree of at least that diameter. If he used three planks, he needed a tree that was at least ten inches in diameter. In the Carpathian Mountains and most parts of Europe to the north and west, forests were common so there was no problem obtaining trees of sufficient size. On the Black Sea plain, however, the cooler and drier climatic conditions that set in around 5000 B.C.E. had, by 4000 B.C.E., initiated a slow shift from forest to prairie. Finding trees large enough for solid-wheel construction may not yet have been difficult in 3000 B.C.E., when the practice of burying wagons with their owners began, but big trees must have been much harder to find by 2000 B.C.E., when vehicles with spoked wheels began to appear.

Two technical explanations, as opposed to a dream of building war chariots, can be considered for the changeover to spoked wheels. One holds that vehicle builders wanted lightweight wheels in order to improve the carrying efficiency of their carts and wagons; the other regards spokes as a response to a dwindling supply of large trees. The archaeological discovery in Europe of wheels made with crossbars instead of from solid wood, or with holes in them, usually in the form of semicircular cutouts, supports the first theory (figure 45). But northern

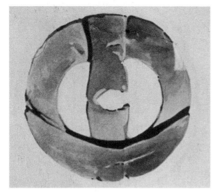

FIGURE 45 Bronze Age wheels with crossbars and semicircular cutouts found in northern Italy. (Institution of Mechanical Engineers, www.imeche.org)

Europe had forests, while southeastern Europe, with its expanding grasslands on the Black Sea plain, did not. Nor did the lands farther to the east in Central Asia up to the northwestern frontier of China. There, the crafting of vehicles using smaller pieces of wood may have become an urgent matter.

Although the impact of war chariots racing along on spoked wheels has captured the imagination of both historians and non-historians, there was a practical benefit to using spoked wheels that was particular to the steppes. Wheels with spokes were not limited in size by the diameter of trees. Like the example shown in figure 44, they could have a diameter of five feet or more instead of thirty inches. This would have greatly facilitated pulling wagons through tall grasses, as we know from the history of American pioneers crossing the Great Plains in covered wagons. The rear wheels of those wagons were usually five or six feet in diameter.

Spoked wheels and two-wheeled carts came on the scene at roughly the same time, but the two developments were not necessarily related. Spokes made possible lighter wheels and larger wheels, but spoked wheels were much harder to build than solid wheels. A villager could craft two solid wheels and fix them to the ends of an axle without special knowledge or training. Solid wheels that rotated on the ends of an axle called for somewhat greater skill, but they still would not have required a specialist to construct them. Spoked wheels, however, called for exact spacing, carefully calculated angles for the insertion of spokes into hub and felloe, and precision fitting of the felloes to ensure a perfect circle. In other words, the specialized craft of the wheelwright probably began with the invention of the spoke. Nevertheless, villagers who did not have access to a wheelwright's skills continued to make solid wheels for their ox-carts into the twentieth century.

Regardless of the design of the wheel, most wheel users after 2000 B.C.E. opted for two-wheeled carts instead of four-wheeled wagons. Four wheels had met the need for mine-cars pushed by hand and for mobile homes pulled at a snail's pace by a yoke of oxen. But the everyday hauling chores in an agricultural economy called for greater maneuverability, as did carrying freight across country. David Anthony and many other historians take it for granted that as soon as wheeled vehicles became available, they were used for all manner of day-to-day tasks. But four-wheeled wagons that could not turn corners could not easily have been so employed. It is more likely that wheeled vehicles were not used for chores like bringing in crops, hauling manure, and carrying trade goods until carts replaced wagons.

Over time, easily maneuverable two-wheeled carts became the common vehicles in most of the world, even when they continued to incorporate the old wheelset technology. Outside Europe and Central Asia, four-wheeled wagons were rarely seen between 500 C.E. and the beginning of the imperialist era, when European settlers and overlords introduced them into their colonies or reintroduced them, as happened in parts of the Middle East where they had existed in ancient times but then vanished.

Wheel-borne domiciles continued to traverse the Black Sea plain, and they spread to adjoining steppe regions from central Hungary to Central Asia. Continuing the customs of the people who had been interred with their wagons in Ukraine and Russia, wagon nomadism became a distinctive feature of life on the steppes. No nomads anywhere else in the world customarily put their homes on wheels. Migratory pastoralists and hunter-gatherers in Africa and the Americas either lived in temporary dwellings that they could simply abandon or disassembled their huts and tepees and loaded the components—sticks, hides, tenting—onto the backs of their animals, their own backs, or travois.

Some steppe homes continued to be built with four wheels, but increasingly ways were found to mount them on two-wheeled carts. In the fifth century B.C.E., the Greek playwright Aeschylus described the homes of Indo-European nomads on the Black Sea plain:

> The wandering Scythians thou shalt find, who dwell
> 'Neath wicker roofs, high-set, on wains [wagons] well-wheeled,
> With bows equipped, death-dealing from afar[2]

His younger contemporary Herodotus agreed, saying of the Scythians: "For when men have no established cities or forts, but are all nomads and mounted archers, not living by tilling the soil but by raising cattle and carrying their dwellings on wagons, how can they not be invincible and unapproachable?"[3] Clay models of Scythian mobile dwellings show both two- and four-wheeled designs, but no spokes (figure 46). The homes were pulled by slow-moving oxen, so they did not have to be light and agile. How long the nomads continued to use solid wheels before climate change caused the big trees to disappear is hard to determine from the available evidence.

The Sarmatians, an Indo-European people related to the Scythians who succeeded them on the steppes, were labeled *hamaksoikoi* (wagon-dwellers) by the Greek historian Strabo in the first century C.E.; and the historian Ammianus Marcellinus, who lived some four centuries after

FIGURE 46 Clay models of Scythian mobile homes, first millennium B.C.E.
(Photograph in author's collection)

Strabo, described the Alans, who were closely related to and succeeded the Sarmatians:

> [T]hey have no huts and care nothing for using the plowshare, but they live upon flesh and an abundance of milk, and dwell in wagons, which they cover with rounded canopies of bark and drive over the boundless wastes. And when they come to a place rich in grass, they place their carts in a circle and feed like wild beasts. As soon as the fodder is used up, they place their cities, as we might call them, on the wagons and so convey them: in the wagons the males have intercourse with the women, and in the wagons their babes are born and reared; wagons form their permanent dwellings, and wherever they come, that place they look upon as their natural home.[4]

HOME ON THE RANGE

All the languages spoken by the nomadic Scythians, Sarmatians, and Alans in the first millennia B.C.E. and C.E. belonged to the Iranian branch of the Indo-European language family. But the speakers of Turkic languages, who succeeded the Indo-Europeans as the primary steppe nomads from the seventh century C.E. onward, readily adopted the practice of wagon dwelling, as did Genghis Khan's Mongols, who were linguistically related to the Turks, when they united the steppes in a great empire in the thirteenth century. Here is how William of Rubruck, a Franciscan friar who journeyed from 1253 to 1255 to the court of the Great Khan Möngke, a grandson of Genghis Khan, described Mongol living arrangements:

> The married women make themselves very fine wagons. . . . One rich [Mongol] or Tartar has easily a hundred or two hundred such wagons with chests. Baatu has twenty-six wives, each of whom has a large dwelling, not counting the other, smaller ones placed behind the large one, which are chambers, as it were, where the maids live: to each of these dwellings belong a good two hundred wagons. When they unload the dwellings, the chief wife pitches her residence at the westernmost end, and the others follow according to rank. . . . Hence the court of one wealthy [Mongol commander] will have the appearance of a large town, though there will be very few males in it. . . . One woman will drive twenty or thirty wagons, since the terrain is level. The ox- or camel-wagons are lashed together in sequence, and the woman will sit at the front driving the ox, while all the rest follow at the same pace. If at some point the going happens to become difficult, they untie them and take them through one at a time. For they move slowly, at the pace at which a sheep or an ox can walk.[5]

This way of life continued on the steppes down into modern times. In 1793/1794, the German zoologist and botanist Peter Simon Pallas made a watercolor painting of Noghay nomads encamped on the Black Sea plain (figure 47). Pallas had resided in Russia under the patronage of

FIGURE 47 Christian Geissler's engraving, after Peter Simon Pallas's watercolor, of a Noghay encampment, 1793–1794. Geissler was a member of Pallas's expedition to southern Russia. (Courtesy of the Qaraqalpaqs, www.qaraqalpaq.com)

Catherine the Great for three decades and observed with a scientist's precision. The Noghays were descendants of the Mongols of the Golden Horde and in Pallas's time were being harried by a czarist state intent on forcing nomads to settle. So his was the last firsthand visual recording of how nomads camped in the region that had given rise to the practice of wagon nomadism almost five thousand years earlier.

Pallas's painting shows a number of round dwellings, called *yurt* in Turkish and *ger* in Mongolian. The white ones are mounted on ox-drawn carts with spoked wheels, though the one in the foreground has been removed and placed on the ground next to the wheeled platform that carried it. The larger, darker *yurt* beside it has belting around it of the sort that came into use to support the collapsible latticework that formed the dwelling's interior walls once portable *yurts* were developed that could be disassembled and carried on carts or by pack animals. Each of the white dwellings still on wheels heads a line of carts with hooped roofs. Baggage carts of this design date back at least to Scythian times (see figure 46) and are explicitly described by William of Rubruck. The unroofed cart at the end of the row in the middle range of the picture probably carries the disassembled parts of one of the larger *yurts*. The settled camp in the background has three white *yurts* and eight of the larger kind.

Although Pallas's painting brings to an end the story of the wagon nomadism on the steppes of Eurasia, a vestige of this tradition may possibly survive in the funeral practices of the Roma, or Gypsies. The ancestors of the Roma migrated from what is now Pakistan to what is now Iraq in the seventh century C.E. Medieval Arabic sources call them Zutt, presumably from the South Asian ethnic name Jat. From Iraq, some groups found their way into central and northern Europe, and others moved into Spain.

The Arab writers make no mention of the Zutt living in wagons, and wheeled vehicles had disappeared from the Middle East by the time they arrived. So it is likely that the nomadic Roma of eastern Europe took up living in wagons, or caravans, after they encountered that lifestyle in Romania, Ukraine, or Hungary. Their cousins who migrated to Spain seem not to have adopted the practice. Speculative though it might be, one wonders whether the Roma tradition of burning the caravan, or sometimes even the automobile, of a deceased person has roots in the wagon burials of the third millennium B.C.E. The Roma say that death taints the vehicle, but also that it might be needed in the afterlife.[6]

The invention of the wheelset and the independently rotating wheel in Europe led to the use, respectively, of four-wheeled cars to remove copper ore from mines and four-wheeled wagons to house and transport nomads on the steppes. The history of the wheel in the Middle East followed a different, not so practical, trajectory that reached its apex with the war chariot—the most written about and glorified of all ancient vehicles.

five

Wheels for Show

The archaeologist Stuart Piggott was cited in chapter 3 for his theory of a Mesopotamian origin of the wheel based ultimately on "inherent historical and technological probability in assuming a more likely initial invention as a part of the complex innovations of this period [ca. 3000 B.C.E.] in the Near East rather than in the simpler complex of Neolithic Europe."[1] Although seemingly convincing, his reasoning is deeply flawed. It conceives of wheels solely as indicators of societal sophistication without taking into account whether the ancient Near East had a particular need for carts and wagons. To be sure, the pictorial and material evidence for early wheeled vehicles in the river valley civilization of Mesopotamia lends some support to Piggott's argument, but the total absence of parallel evidence from the contemporary and equally sophisticated river valley civilization of Egypt raises a red flag. If wheeled transport existed in only one of these two societies between 3000 and 1300 B.C.E., but the two cultures are known to have been in contact and shared in "the complex innovations of this period in the Near East," then the premise of Piggott's argument falls.

To explain why Mesopotamia had wheels and Egypt did not, I propose that the wheels in Mesopotamia originated for showing off the grandeur of kings and priests rather than for general utility. This does not mean that there were no royal show-offs in Egypt, of course. The pharaohs, after all, commanded the building of pyramids and temples that were far more grandiose than those in Mesopotamia. But the broad and peaceful Nile River provided a magnificent avenue for royal boats and barges, which were sometimes buried with a pharaoh, while torrential springtime floods rendered the Tigris and Euphrates Rivers in Mesopotamia too turbulent and unreliable for royal regattas. Even today, excursion boats ply the Nile by the score, but the same cannot be said of the Tigris and Euphrates in (peacetime) Iraq.

In discussing mine-cars in the Carpathian Mountains and wagon-borne dwellings on the Black Sea plain, I have focused on a single question: What transport needs developed in those two regions that could not be served by the age-old practice of dividing loads into manageable size and putting them on the backs of animals and humans? When that question is posed for Mesopotamia, no answer presents itself. Indeed, few parts of the world had less need for wheeled vehicles than did Mesopotamia in the fourth millennium B.C.E. Although the Tigris, the Euphrates, and some of the wider irrigation canals dug parallel to them were not suitable for royal yachting, they worked perfectly well for the small boats that performed day-to-day hauling chores. This is why Uruk, Ur, Lagash, and other Sumerian cities of the third millennium B.C.E. were normally located near riverbanks, though not so close as to be endangered by the annual floods.

The farming population that supplied the cities' needs lived in villages situated on or near irrigation canals. The materials used to build everything, from the giant pyramidal ziggurats at the center of the cities to the houses of villagers, consisted of sun-dried or baked bricks, reeds, and slender poles, all of which could easily be divided into loads suitable for carrying by pack animals. And the date palm, the only tall, straight tree that grew in abundance, was not only ill-suited for sawing into planks, but also too important as a food source to be casually cut down and used for lumber.

So what does the surviving evidence, consisting of a few pictograms and fragments of clay models, suggest as the reason for the appearance of wheeled transport in an early phase of Sumerian civilization, dating from about 3100 to 2900 B.C.E.? The pictograms show a peaked-roof enclosure on a platform with a turned-up front. Some of the platforms are sledges, which slid over the ground on solid wood runners. Others were drawn with two black circles underneath them (figure 48). One

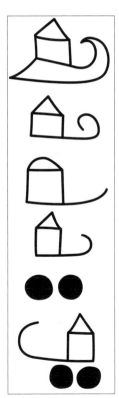

FIGURE 48 Pictographs of early Sumerian sledges, with and without possible wheels.

specialist on the history of the wheel, László Tarr, maintains that the pictograms prove "that the wheeled vehicle was a developed and widely used transport device in the Jemdet Nasr [that is, early Sumerian] period, since otherwise no special pictographic character would have been used to designate it."[2]

But what about the images of sledges without the black circles? Do they also illustrate a widely used transport device? Tarr does not comment on the absence of anything connecting the circles to the bottom of the sledge or the absence of any means of propulsion. Other scholars look at these peculiarities and conclude that the circles represent the ends of rollers rather than wheels. But they, in turn, do not comment on the number of circles: there are never more than two. Yet if they depict rollers, there should be at least three. Since rollers are not attached to the bottom of a platform supporting a load that it is being used to move, when the load moves forward, a new roller, which may be a back one that is no longer under the load, has to be placed under the front end of the platform to keep it from falling to the ground. Thus at any given moment, the load should be balanced on at least three rollers.

Even more nonsensical, and not just because only one roller is shown, is a frequently reproduced drawing that imagines how a roller being used to move a huge stone may have evolved into wheels (figure 49). The idea of wheels evolving from rollers goes back to at least 1881, when E. B. Tylor, one of the founding fathers of anthropology, wrote:

> Though the origin of the wheel-carriage is even more totally lost in prehistoric antiquity than that of the plow, there seems nothing to object to the ordinary theoretical explanation . . . that the first vehicle was a sledge dragged along the ground; that, when heavy masses had to be moved, rollers were put under the sledge, and that these rollers passed into wheels, forming part of the carriage itself. . . . If, now, the middle part of the trunk of a tree used as a roller were cut down to a mere axle, the two ends remaining as solid drums, and stops were fixed under the sledge to prevent the axle from running away, the result would be the rudest imaginable cart. *I am not aware that this can be traced anywhere in*

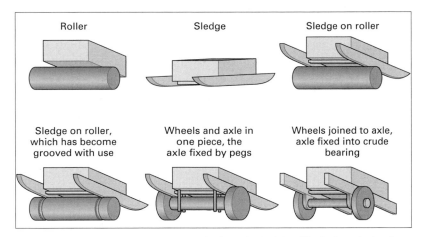

FIGURE 49 The hypothetical, but illogical, evolution from
sledge to roller to wheel.

actual existence, either in ancient or modern times; if found, it would be of
much interest as vouching for this particular stage of invention of the
wheel-carriage.[3]

It is not surprising that no evidence of this transition "can be traced
anywhere in actual existence" because the theory itself makes no sense.
Leaving aside the facts that the process Tylor outlines would result in a
wheelset, rather than independently rotating wheels, and that archaeologists have never found evidence of wheelset vehicles south of Turkey,
the historical reality is that Egypt, where enormous stones were moved
in great quantity, did not have wheels, but southern Mesopotamia, which
did have wheels, is an alluvial plain entirely devoid of building stone.
Instead of stone, the early Sumerians constructed their monumental
buildings out of bricks, which could easily be transported in manageable loads.

And even if they had used huge building stones, the Sumerians would
have discovered that the process imagined by Tylor would have produced

a weaker rather than a stronger transport device. The point of putting rollers underneath a heavy load, as shown in figure 49, is to distribute the weight of the load over a long length of wood. The same is true even if sledge runners are placed between the load and the rollers because the contact the runners make with the rollers at two specific points will still spread the weight of the load over the entire length of the rollers. The sledge runners might indeed wear grooves in a roller, as the figure illustrates, but cutting away the wood between the grooves to convert most of the roller into an axle while leaving large wheel-like ends would entirely defeat the purpose of the roller. Instead of a heavy weight being distributed over the length of the roller, the weight would be borne solely by the solid wheel-like drums remaining outside the grooves worn by the runners.

Ridiculous in principle, this scheme makes even less sense for lower Mesopotamia in early Sumerian times. As already mentioned, the date palm was the only tall tree abundantly available. Unlike the oaks and other hardwood trees that provided the wood used to construct the early wheels discovered in Europe, however, palm trees do not have hard heartwood cores buried within softer outer layers of sapwood. Their trunks are more like flexible, fibrous plant stems and do not have annual rings. The hardest layer is on the outside, and the middle is much softer. Therefore, a date palm trunk used as a roller would become weaker and weaker at the points where the sledge runners wore grooves, and an axle made entirely of the innermost part of the trunk would be extremely weak.

Thus I conclude that, contrary to Piggott's theory, the wheel was not invented in Mesopotamia, where the earliest known pictograms and broken clay models are a century and more younger than the Borenáz clay models, and it certainly did not evolve there from the use of rollers. Yet the pictograms of wheeled sledges and the fragments of models still have to be explained. So following the type of questioning pursued in regard to the Carpathian Mountains and the Black Sea plain, I would ask: What new transport needs arose between 3100 and 2900 B.C.E. in lower Mesopotamia?

As Piggott has pointed out, the Near East was the first part of the world to witness the development of sophisticated urban civilizations dominated by priests and kings who monopolized the society's resources. I propose, therefore, that even though in Mesopotamia this process did not require the transport of giant building stones, it did create a need to display the grandeur of each city's gods and overlords; and this, in turn, created opportunities to use vehicles, whether on sledge runners or equipped with wheels, in processions and parades.

Archaeological investigations have shown that village life in the Carpathians and on the Black Sea plain in the fourth millennium B.C.E. was basically egalitarian. Kings and priests may well have cherished a few costly possessions, but there is no indication that they designed their communities around monumental buildings and grand processional boulevards. Not so in Mesopotamia. Priests and kings were sometimes rivals for power, but together they completely dominated both Sumer and the societies that succeeded it. Their power became manifest in grandiose buildings, extravagant burials, and glittering luxury items. Nothing comparable has been discovered from that time period in Europe.

Let us look, then, at the early Sumerian pictograms with a view to how they might relate to a taste for ostentatious display. A square topped by a triangle sits at the back end of a platform that has a turned-up front (see figure 48). This appears to be a sledge, and a somewhat later image shows just such a sledge, without wheels, being pulled by oxen. Another pictogram, also somewhat later, shows the profile of a person sitting inside the square, though with no indication that the person is driving the oxen. While it is barely conceivable that some real person fancied squeezing into a tiny roofed room in one of the hottest parts of the world so that oxen could pull him from place to place on a sledge at a pace slower than he could walk, it is more likely that the square represents a portable shrine and the sledge is a way of parading it. In other words, the little room shelters the statue of a god that is being pulled on a sledge, with or without the help of wheels, as part of a religious procession.

If this surmise is correct, then the use of wheels in early Sumerian times, when temple priests dominated the common folk in each city-state, had more to do with display and ritual than with economic utility. Since the pictograms never show any load other than the little square room, they speak to a highly specialized use of wheels rather than to what Tarr termed a "developed and widely used transport device."

Although counterintuitive, given the common assumption that wheels were invented because they were economically efficient, this theory is not difficult to explain. Religious and ceremonial processions around the world often feature shrines, statues, urns, and caskets carried on the shoulders of devotees. Such parades bespeak an urban lifestyle and highly organized religious and political institutions, both cardinal features of Sumerian Mesopotamia, rather than the social organization of hunter-gatherers, herders, or village-based farmers. Processions have features that would have accommodated the limitations of vehicles with four solid wheels, which are difficult to steer. First, processions are slow, as parades still are. Onlookers do not want their object of reverence to pass by too quickly. Second, the distances covered are short compared with what would be involved in transporting goods and people around town or across country. They typically involve movement to or from a sacred structure in a city or to a place of burial or cremation. Third, the shortness of the processional route makes it possible for it to be leveled, if not paved. The road may even be reserved exclusively for ceremonial purposes. Fourth, the size and weight of the object being borne is not limited by considerations of efficiency (figure 50). And fifth, the cost of the processional vehicle, whether calculated in terms of money or of the time and energy devoted to its construction, is of minimal importance. Indeed, the more costly the procession, the better because the purpose of the ceremony is to impress devoted viewers and participants.

The wheels that are used on processional shrines, floats, and funeral vehicles may differ greatly in size, number, and construction from those that are used on more utilitarian carts and wagons. Take, for example, the giant Indian shrines known as juggernauts, which were once so heavy and unstoppable that they crushed devotees who did not get out of the way. Some of them had many wheels, each on its own axle (figure 51).

FIGURE 50 Relief of a four-wheeled
Roman processional vehicle, with a statue
of a god in a peaked-roof square chamber,
drawn by elephants.
(Photograph © The Trustees of the
British Museum / Art Resource, NY)

As Ezra M. Stratton, an eminent nineteenth-century authority on carriages, relates:

> [T]hen comes the grand procession. The car is twenty feet high, constructed like a pyramid, and is twenty feet square. It is mounted on twenty-four wheels, each wheel four feet in diameter and more than a foot thick. These wheels are arranged in three rows, eight wheels in a row, and placed two feet apart, so that whoever falls under them is crushed. . . . [A] hundred thousand people struggle with each other for the privilege to draw the "infernal machine."[4]

FIGURE 51 A fanciful design of an Indian juggernaut with sixteen wheels, as shown in the watercolor *The Idol Juggernaut on His Car During the Rath Jatra* (ca. 1820–1822). (© Victoria and Albert Museum, London)

Two-wheeled carts, though favored by farmers and haulers, seldom appear in processions, with the exception of war chariots such as the one used in the earliest Roman kingdom when a triumphant general would be painted vermilion like a statue of Jupiter and would parade through the city in a chariot as the avatar of the victorious god. Four-wheeled vehicles did a much better job of keeping a ceremonial object like a statue or an urn level, particularly over rough surfaces. This explains why four or more wheels persisted for processional purposes in Mesopotamia and elsewhere, including northern Europe (figure 52), even after the two-wheeled cart became the vehicle of choice for everyday transport.

The wagon burials on the Black Sea plain certainly involved funeral processions, but the vehicles were solidly practical and unornamented, unlike what would be expected of conveyances built solely to be ceremonially interred. In the Sumerian city-states, however, priests wanted to serve their gods by displaying them, and the kings who arose as rivals to the priests felt the same way about displaying their own royal persons. An ornamented box found in a Sumerian burial epitomizes such kingly display. It is known as the Standard of Ur (figure 53). The monarch, who died in 2550 B.C.E., stands in the center of the top tier of figures with his unoccupied wagon behind him, while the bottom tier shows four battle wagons, each with a warrior standing behind a driver, running over the bodies of slain enemies. The wagons in the battle scene, unlike the royal vehicle above it, are stocked with javelins. Each wagon is drawn by four long-eared animals that look like horses, the same sort of animal that appears on a bronze statuette of roughly the same period showing a single rider straddling the framework of a small two-wheeled cart (figure 54). Both the four-wheeled and two-wheeled

FIGURE 52 The Trundholm sun chariot, a model of a six-wheeled processional vehicle bearing a bronze disk and dating to 1800 to 1600 B.C.E. that was found in Denmark. (National Museum, Copenhagen. Album / Art Resource, NY)

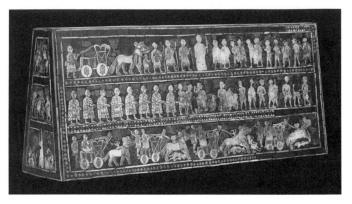

FIGURE 53 Battle wagons depicted on the Standard of Ur from southern Iraq, ca. 2600–2400 B.C.E. The monarch stands in the middle of the top row of the "War" panel, which is worked in inlaid mosaic. (Photograph © The Trustees of the British Museum / Art Resource, NY)

FIGURE 54 A figurine of a Sumerian straddling the saddle of a two-wheeled cart drawn by onagers. (Photograph in author's collection)

vehicles show up in other artifacts from the Sumerian period as well, the latter in the form of miniature clay cart-frames with axle holes and either a saddle for a standing driver to clasp between his knees for stability or a seat with a back.

The most striking characteristic of these vehicle designs is their limited usefulness. Vivid imaginings of how the war wagons functioned in battle abound, but the most hard-headed specialists on Mesopotamian wheels and harnessing, Mary Littauer and Joost H. Crouwel, are skeptical:

> The Sumerian wagon afforded hardly any protection; not only its crew, but its motive power were highly vulnerable, and its field of operation was strictly limited to level and open ground. In addition, without a swiveling front axle, it could only be turned in a wide arc or by manual lifting or levering of its rear wheels. This limitation and that caused by the very elementary control over the team, greatly restricted its maneuverability.

They nevertheless conceded that, despite the limitations of the wagons, they do seem to have been present at actual battles:

> These wagons could have functioned as mobile arsenals and firing platforms from which javelins, carried in sheaths attached to the right front corner of the breast work, could have been thrown most effectively when moving along the front or flanks of massed infantry. . . . The vehicles could also have been used to convey the king and important officers to the field of battle.[5]

As for the two-wheeled carts that show a driver straddling a saddle, they are so narrow that it is hard to imagine them having any useful purpose other than showing off the rider or perhaps racing another cart.

To return to a point made by both David Anthony and Stuart Piggott, it is generally assumed—indeed, taken as gospel by most people who ponder the invention of the wheel—that the use of wheels spread rapidly once the basic discovery was made. But how does this assumption play out in Mesopotamia? The first wheels in Europe precede the earliest Sumerian pictograms, but the earliest actual depictions of Mesopotamian vehicles, such as those on the Standard of Ur, bear no resemblance to the designs seen either in the Carpathian Mountains or on the Black Sea plain. Not only did the Sumerians not use their wheeled vehicles for the same purposes as the Indo-Europeans, at least at the outset, but they constructed their wheels differently and used animals other than oxen to pull at least some of their wagons and carts. Moreover, no piece of physical evidence tying the early Sumerian vehicles to either the Carpathian region or the Black Sea plain has ever been found in lower Mesopotamia.

It seems likely, therefore, that the wheel spread as an idea but not as a technology. The spread of paper from China, where it was made from tree leaves, to the Islamic Middle East, where it was made from rags, affords a plausible comparison. It is not hard to imagine travelers or traders from Mesopotamia returning home from trips into Europe with descriptions of vehicles they had seen but with no precise knowledge of how they were built. This scenario seems more probable than that Mesopotamian craftsmen, after tens of thousands of years without the wheel, just happened to hit on the idea independently only a century or two after the wheel was invented some 1500 miles away in eastern Europe.

The case for or against independent invention hinges on the importance assigned to the design of the wheels and the harnessing of the animals. The Standard of Ur (figure 55) and a limestone relief, also from Ur, showing a two-wheeled cart designed to be straddled by its driver (figure 56), detail the wheels' structure quite clearly. The wheels rotate freely. Their main body, while solid, is composed of two pieces of wood shaped like quarter-moons on either side of a football-shaped central piece, which sometimes extends all the way to the rim. Short battens help hold the three central pieces together. A nave made from a different piece of wood protrudes from the center, and a linchpin goes through the end of the axle, which pokes through the nave.

Although many solid wheels in Europe were made from three pieces of wood, the planks usually were straight and parallel and extended

FIGURE 55 Detail of the "War" panel of the Standard of Ur, showing the structure of the wheels of a four-wheeled battle wagon. (Werner Forman / Art Resource, NY)

FIGURE 56 Relief of a two-wheeled Sumerian riding cart, showing the structure of the wheels. (University of Pennsylvania Museum of Archaeology and Anthropology, Philadelphia)

to the wheel's circumference, and there is no comparable use of the quarter-moon shape (see figure 32). Moreover, the Mesopotamian wheels, unlike their contemporary European counterparts, show a rim circling the entire assembly. The rim is particularly clear on the limestone relief (see figure 56). It may have been bronze, a hard alloy of copper and tin then coming into use in Mesopotamia, but it more likely was thick leather. Other images reveal fewer structural details but sometimes show a knobbed rim that apparently was composed of the rounded heads of large bronze nails (see figure 54), which would have functioned like cleats on an athletic shoe. Whether a solid band or a series of nail heads, strengthened rims would have reduced the wear on the wheel and helped hold its parts together. In sum, these Sumerian wheels from Ur were both sophisticated and expensive. But these may not have been the traits of the wheeled sledges from five hundred years earlier, which show up only as black circles on pictograms. In other words, Sumer may have started with solid wheels, and then evolved the more elaborate forms shown on the Standard of Ur in isolation from developments in Europe.

The question of animal harnessing relates not just to the independence from Europe of the Mesopotamian tradition, but also to an eventual cross-fertilization of Mesopotamian and Black Sea practices in the era of the chariot (chapter 6). The import of this issue requires a digression on the subject of animal domestication, however, particularly the domestication of the onager, which is also termed a hemione, or half-ass (from the Greek *hemi-* [half] and *onos* [ass]). Vast herds of wild horses roamed the regions penetrated by the wagon nomads trekking eastward from the Black Sea plain. But there were no wild horses in Mesopotamia. The local cousin of the horse there was the onager, which is bigger than a donkey (a native of North Africa not known in Mesopotamia in early Sumerian times) and looks like a horse with long ears. These are the animals that are shown pulling the battle wagons on the Standard of Ur and the carts with drivers straddling a saddle. However, there are no surviving indications that onagers played any role at all as domestic animals in Mesopotamia in later times. Donkeys and horses eventually replaced them.

Historians of animal domestication find it difficult to account for the onager pulling Sumerian vehicles because that would make it the only animal known to have been used for a domestic purpose and then to have fallen out of domestic use and to survive only in the wild. Jared Diamond, who ascribes world historical importance to the luck of certain peoples in living in proximity to wild animals that are suitable, in Diamond's view, for domestication, observes: "Some ancient depictions of horselike animals used for riding or for pulling carts may refer to onagers. However, all writers about them, from Romans to modern zookeepers, decry their irascible temper and their nasty habit of biting people. As a result, although similar in other respects to ancestral donkeys, onagers have never been domesticated."[6] Absurd as it is to imagine that the Sumerians were able to get four animals that were too ornery to be domesticated to pull together under harness, it is important to Diamond's thesis that only a handful of wild animals were suitable for domestication. Hence he values the opinions of later Roman writers and modern zookeepers over the clear pictorial evidence from Mesopotamia.

The more plausible solution to the puzzle of the disappearing domestic onagers is to assume that in Sumerian times, when onagers were more abundant, they did not have so pronounced a fight-or-flight mentality as they do today, when their few surviving descendants teeter on the brink of extinction. Generally speaking, domestic animals cease to be utilized for specific functions once a more efficient species is domesticated or becomes available. Mares were traditionally valued for milk in Central Asian societies that raised few cows, for example, but horse milk did not continue to be drunk in Europe, where cows were plentiful and more productive. In similar fashion, domestic onagers did not retain their value as draft animals once donkey herders migrated into Mesopotamia from the west (ca. 2000 B.C.E.) and horse herders occupied the mountains that border the Tigris valley on the east (ca. 1500 B.C.E.). As far as can be determined, onager herding never became a focus of pastoral life, as did donkey and horse breeding. Hence the cost of a domestic onager was probably higher than that of its equine relatives.

Although originally pulled by oxen, as indicated by the early Sumerian image of two oxen pulling a shrine mounted on a sledge, wheeled vehicles stimulated the Mesopotamians' interest in finding faster draft animals. This happened elsewhere as well. Experiments with new domestic animals took place independently of what was happening in Mesopotamia when the wagon nomads of the Black Sea plain entered the range of the horse and the two-humped camel east of the Caspian Sea, probably in Kazakhstan. By 2500 B.C.E., Indo-Iranian nomads were occasionally using camels instead of oxen to pull their mobile homes in the region of modern Turkmenistan (figure 57).

The wagon nomads' encounter with horses was more fateful than that with two-humped camels, however. David Anthony maintains that

FIGURE 57 An Indo-Iranian camel wagon from Turkmenistan, ca. 2500 B.C.E.
(Photograph in author's collection)

WHEELS FOR SHOW

people learned to ride horses long before they tried hitching them to carts, which may explain why the method of harnessing horses by means of a bit in the mouth differed from that of attaching reins to the nose rings used for oxen, onagers, and two-humped camels. The bit provides a more precise degree of control than the nose ring because the mouth of an animal is more sensitive than the flesh of its nostril, and precise control is what was called for when two-wheeled carts evolved into war chariots, which had to be fast and maneuverable. This momentous development took place in Kazakhstan around 2000 B.C.E., some five hundred years after the Sumerians began to harness onagers to their carts and wagons.

The development of the war chariot opened a new period in the history of wheeled vehicles in Mesopotamia. But even though the wheel was not invented there, the evidence from the pre-chariot era in Sumer warrants attention both for the ways in which it differs from the evidence for wheel use in the Carpathian Mountains and on the Black Sea plain, and for two patterns of use that seem to have originated in Mesopotamia and continued into the chariot era.

First, royal and priestly processions, ostentatious display, and precarious riding behind a quartet of speeding onagers were hallmarks of the Sumerians' experience of wheels. The same cannot be said of the miners of the Carpathian region or the nomads of the Black Sea plain. Historians commonly speculate on the ways in which wheeled transport transformed the economic life of ancient societies. But showing off and reveling in speed could be as great a goad to invention as hauling bricks and manure.

Second is the likelihood that Sumerian drivers straddling their saddles behind four-onager teams inaugurated one of the most energizing practices in the entire history of the wheel: competitive racing. No concrete evidence survives, but it is hard to imagine that they did not race their vehicles. Why else use four fleet-footed animals to pull a single rider?

Chariot racing, which was to persist for 2500 years in the West, but did not spread to East Asia and South Asia, thus seems to represent a fusion between the Sumerian need for speed, on the one hand, and, on the other, technical innovations like the lightweight spoked wheel being pioneered in Kazakhstan.

six

The Rise and Demise
of the Charioteer

hree birthplaces, three ways of thinking about wheels.

First, in the Carpathian Mountains, miners pushed four-wheeled ore-cars along stone tunnels. Wheels and axles rotated together, and European mine-cars continued to be designed with wheelsets for five thousand years before emerging into the light of day as railroads.

Second, on the Black Sea plain and to its east, ox-drawn wagons trundled across the steppes pulling the dwellings of nomads. Their solid, thick-naved wheels rotated independently of each other at the ends of axles.

Third, in Sumer, the awe-inspired faithful gazed at religious shrines mounted on ox-drawn sledges, with and without wheels, while warrior overlords paraded in grandiose but clumsy battle wagons and wrangled half-wild onagers in perilous dashes across the desert.

Archaeological evidence indicates that these scenes evolved chronologically in the sequence in which they have been described, but surely there were interconnections. Not direct copying perhaps, since the wheels themselves and the uses to which they were put differed so greatly, but at least a trickle of reports by travelers on what they encountered on their journeys.

The technological divergence between Chinese and European wheel-barrows offers a comparison with that among the wheel cultures of the Carpathians, the Black Sea plain, and Mesopotamia. Exchanges of information and goods along the Silk Road—which linked China, the Middle East, and Europe—date to at least 300 B.C.E., which is roughly the time of invention of the Chinese wheelbarrow. They continued for almost two thousand years. At the European end of the Silk Road, there are hints that a vehicle similar to the wheelbarrow may have been used in classical Greece as early as its counterpart in China; if so, it did not catch on. Convincing visual evidence of wheelbarrow use does not appear before the fall of the Roman Empire in the fifth century C.E. In the Middle East, which had more consistent contact with lands along the Silk Road than did Europe, wheelbarrows never came into use.

Travelers from the West surely set eyes on Chinese wheelbarrows, which became widely popular for moving both people and goods. So it is hard to imagine that they never reported on what they had witnessed when they returned home. Yet tales of a one-wheeled vehicle seem to have fallen on deaf ears. Perhaps people who heard about it found the notion of men doing the work of oxen repugnant. Perhaps the travelers' reports did not explain how the Chinese managed to balance great loads over single wheels. In any case, when Europeans finally did devise single-wheeled vehicles, they produced distinctly inferior designs: small-diameter wheels that could not roll on bumpy surfaces, a load distribution that required the operator to lift as well as push, and no capability of cross-country travel. As was observed earlier when discussing wheels in the Western Hemisphere, *simply thinking of or hearing about a wheeled conveyance was not enough to call it into being.* There had to be a rationale for its use, and local artisans had to call on the skills they knew to devise ways of responding to that rationale.

Unlike those from along the 5000-mile Silk Road, travelers' reports from within the closely interconnected geography of southeastern Europe, the Black Sea plain, and Mesopotamia—a region approximately 1500 miles square and criss-crossed by trade routes—inevitably gave way to more concrete technological interaction. The three original domains

THE RISE AND DEMISE OF THE CHARIOTEER

of wheel use, which were quite separate from one another in 3000 B.C.E., had increasingly overlapped and experienced technical cross-fertilization by 2000 B.C.E. At a general level, and for most purposes, the two-wheeled oxcart with independently rotating wheels became the standard vehicle everywhere. But between 2000 and 1200 B.C.E., a more specialized form of two-wheeled vehicle, the chariot, came to claim center stage because of its impact on the history of war and the heroic posturing of the kings and warriors who rode in it. With the chariot, as with a Ferrari or Lamborghini, it was not just the vehicle that was important. It was what the vehicle said about its driver.

Two images reflect a fusion between the Black Sea tradition and the Mesopotamian, and it is out of that fusion that the chariot emerged. One, a relief, comes from Aleppo in northern Syria. It depicts one of the myriad storm gods revered by the Hittites, an Indo-European people who for the most part lived farther north in what is today central Turkey (figure 58). The god's conveyance is not a war chariot, but it looks like a precursor to one. The wheels of his two-wheeled cart do not have

FIGURE 58 Relief of a Hittite storm god and his bull-drawn cart.
(Photograph in author's collection)

spokes, and it is drawn by a potent bull rather than by a pair of oxen or a team of horses. Moreover, the seat on which the god would ride if he were mounted resembles the sort of saddle that Mesopotamian riders straddled behind their teams of onagers (see figure 54). War chariots were different from the storm god's cart. They had a high front, an open back, and a partial railing along the sides. The god's cart would have offered a warrior no protection in battle.

Although the crossbar construction of the wheel looks more European than Mesopotamian (see figure 45), it is reasonable to see this image as evidence of the extension northward into Turkey of the tradition of godly display described in chapter 5. Yet the Hittite kingdom is recognized as being the first great empire to deploy horse-drawn chariots on the field of battle. At its height under Suppiluliuma I, who ruled between roughly 1344 and 1322 B.C.E., the Hittites provided a model of horse training and chariot fighting that was adopted throughout the Middle East. Egypt, in particular, which had never availed itself of wheeled transport, finally entered the age of the wheel.

The second image appears on a silver bowl found in northern Afghanistan, a land that has been traversed, settled, and fought over by peoples speaking Indo-European languages from at least 1500 B.C.E. (figure 59).[1] Two men ride in an ox-drawn wagon with four wheels. Both the vehicle and the structure of the wheel recall the Mesopotamian battle wagons depicted on the Standard of Ur, but no weapons are shown. A

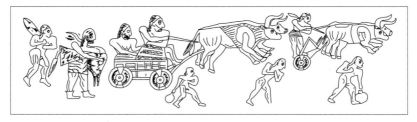

FIGURE 59 A hunting scene, showing both two- and four-wheeled vehicles, that decorates a silver bowl from Afghanistan, perhaps ca. 2000 B.C.E.

THE RISE AND DEMISE OF THE CHARIOTEER

third man stands precariously in a light, two-wheeled cart with similarly designed wheels. It, too, is drawn by oxen. A hook shape slanting upward and forward between the driver's legs suggests in a sketchy manner that he is straddling a Mesopotamian-style saddle. He may be carrying a weapon over his shoulder.

Five men on foot, four naked and one clothed, accompany the two vehicles. One naked man carries a club on his shoulder, while two others hold what might be water bottles. The clothed man, wearing a kilt and boots, is larger than the naked men and carries what could be the headless carcass of an animal (note the tail) under one arm and what could be a bow over the other shoulder. In the absence of armor, a more obvious display of weapons, or fallen enemies, the scene as a whole suggests hunting rather than warfare. Yet the function of the vehicles is obscure. Why lug a heavy carcass rather than putting it in the wagon? Why ride in such a flimsy two-wheeled cart? Why use slow-moving oxen instead of horses, which had surely been domesticated by the time the bowl was crafted?

Just as the Hittite storm god portrayed on the relief as driving a more primitive cart than the chariots deployed by actual Hittite war leaders suggests that the Sumerians' association between divinity and driving a vehicle preceded the development of the horse-drawn war chariot, so the wagon and cart depicted on the silver bowl seem to indicate an expansion far into Indo-European territory of Mesopotamian vehicle designs. In both images, vehicle riding bespeaks prestige more than utility, an attitude that would persist well into the age of chariots.

Light, two-wheeled carts equipped with spoked wheels and drawn by teams of horses, the immediate precursors of the war chariot, first showed up around 2000 B.C.E. at a site called Sintashta. The locale is well to the north of Afghanistan in what is today the frontier between Russia and Kazakhstan, and the people who lived there spoke an Indo-European language. What transformed these carts into war chariots was a fusion of their innovative technology with the Mesopotamian conception of the charioteer as a god-like warrior. The image of the Egyptian pharaoh Ramses II firing arrows while standing in his speeding chariot

FIGURE 60 Relief at Abu Simbel, Egypt, commemorating the victory of Pharaoh
Ramses II at the Battle of Kadesh in 1274 B.C.E., 1264–1244 B.C.E.
(Photograph in author's collection)

at the Battle of Kadesh in 1274 B.C.E. epitomizes this glorification of the
chariot-borne warrior (figure 60). In thinking about famous battles like
Kadesh, many historians, including Mary Littauer and Joost H. Crouwel,
have imagined spears being hurled from chariots at masses of terrified
infantry,[2] but the sketchy surviving descriptions of these clashes scarcely
mention either spears or infantry. As in the picture of Ramses II, the bow
and arrow was the chariot warrior's weapon of choice.

The *Rig Veda*, a collection of hymns composed in an archaic form of
Sanskrit around 1500 B.C.E., in all likelihood in the region of northern
Afghanistan, blesses the chariot warrior and his weapon: "With the bow
let us win cows, with the bow let us win the contest and violent battles
with the bow. The bow ruins the enemy's pleasure; with the bow let us
conquer all the corners of the world."[3] And a passage in the *Mahab-*

THE RISE AND DEMISE OF THE CHARIOTEER

harata, a later Sanskrit epic devoted to a mythic war fought in India during the chariot age, reads: "The two great-spirited and powerful kings struck out at each other, roaring like two bulls in a cowpen. The chariot fighters circled each other on their chariots, loosing arrows as nimbly as clouds let go their water streams."[4] The *Mahabharata* also contains the most famous contemplation of chariot warfare ever penned in the form of the *Bhagavad Gita*, a segment of the poem in which the god Krishna, serving as the charioteer for the hero Arjuna, gives advice and instruction about the duties of warriors. Even when fighting against kinsmen, the god explains, the true warrior must stick to his bloody task.

In practical terms, the bow and arrow, not the spear, was the dominant weapon in chariot battles, and infantrymen were present only in small numbers in the role of runners protecting the flanks of the archers in the chariots. As the battle began, two lines of charioteers would face off on a flat plain and drive at each other, firing arrows. To get a good view of the enemy and have room to maneuver, the chariot lines would be widely spread and only one or two vehicles deep. The chariots might slow as the lines drew near each other, but the line that had put several of the opponent's chariots out of action would then exploit the resulting gap to drive a wedge into the opposing line or, possibly, would split to go around it on one or both sides.

Around 800 B.C.E., however, some five centuries after the Battle of Kadesh and four centuries after the siege of Troy by Achaeans invading from Greece, the poet Homer composed the greatest of all accounts of ancient warfare in his epic *Iliad*. But he does not describe any battles fought from chariots. Homer's heroes ride in chariots to and from the battlefield, and they humiliate fallen enemies by dragging them in the dust behind their vehicles; but actual fighting is done on foot with spear and sword. Not only does Homer never portray archers shooting arrows while driving chariots at full tilt, but he speaks slightingly of warriors who prefer the bow to the spear. Something had changed in the nature of chariot warfare.

Evidently, the centuries that separated the actual siege of Troy from its immortalization in Homer's *Iliad* had witnessed the end of confrontations between opposed lines of archers riding in chariots. Homer knew that the Greeks and the Trojans had once employed chariots, but the memory of how they had used them had been lost. During the intervening centuries, innovative commanders had adopted swarming infantry tactics and taken to strewing obstacles on the battlefield and avoiding flat plains to counter the advantages originally enjoyed by chariot-borne warriors. The attachment of long blades to the sides of later chariots is a sign of a new need to protect against massed infantry.

Yet some Middle Eastern kings were slow to get the message that chariots no longer ruled the battlefield. When the Roman general Lucullus won a battle against Mithridates VI, a monarch in northern Turkey, in the first century B.C.E., a full millennium after the Battle of Kadesh, he seized as part of the spoils ten chariots with yard-long scythes extending from their sides.[5] And Pompey, the general who finally defeated Mithridates VI, rode his defeated foe's gem-studded chariot in a triumphal parade through Rome. One should note, however, that the officers following Pompey rode on horseback or marched on foot.[6] The day of the horse-mounted warrior was at hand.

Much later, in the fourth century C.E., the Roman military writer Vegetius ridiculed the very idea of chariot warfare:

> The armed chariots used in war by Antiochus [III (r. 222–187 B.C.E.)] and Mithridates [VI (r. 120–63 B.C.E.)] at first terrified the Romans, but they afterwards made a jest of them. As a chariot of this sort does not always meet with plain and level ground, the least obstruction stops it. And if one of the horses be either killed or wounded, it falls into the enemy's hands. The Roman soldiers rendered them useless chiefly by the following contrivance: at the instant the engagement began, they strewed the field of battle with caltrops, and the horses that drew the chariots, running full speed on them, were infallibly destroyed. A caltrop is a machine composed of four spikes or points arranged so that in whatever manner it is thrown on the ground, it rests on three and presents the fourth upright.[7]

THE RISE AND DEMISE OF THE CHARIOTEER

In Central Asia, where the chariot was invented, the shift from chariotry to cavalry came quite early. The Scythians, whom we encountered earlier roaming the steppe as wagon nomads, were one of many Central Asian peoples who fought on horseback with bows and arrows. Yet a mythic conception of gods riding in chariots that can already be seen in the relief from Aleppo of a Hittite storm god driving a precursor of the chariot persisted, especially among peoples like the Scythians who spoke Indo-European languages.

Despite their ups and downs as war vehicles, chariots consistently symbolized both earthly nobility and celestial divinity. Homer may not have known how chariot battles had been fought, but he rhapsodizes about a chariot race that highlighted the funeral games staged by the hero Achilles to honor his slain friend and charioteer Patroklos:

> *. . . Rapidly they made their way over the flat land*
> *and presently were far away from the ships. The dust lifting*
> *clung beneath the horses' chests like cloud or a stormwhirl.*
> *Their manes streamed along the blast of the wind, the chariots*
> *rocking now would dip to the earth who fosters so many*
> *and now again would spring up clear of the ground, and the drivers*
> *stood in the chariots, with the spirit beating in each man*
> *with the strain to win, and each was calling aloud upon his own*
> *horses, and the horses flew through the dust of the flat land.*[8]

Images of a cloud or "stormwhirl" roiling beneath the breasts of the horses and of the racing chariots springing clear of the ground, the manes of the horses flying, reinforced the listener's fantasy that the charioteers resembled gods driving heavenly chariots across the sky.

The earliest evidence of wheeled vehicles from the fourth millennium B.C.E., whether clay cups from the Carpathian Mountains or ox-drawn wagons buried on the Black Sea plain, does not suggest a visualization of heavenly affairs. Four-wheeled vehicles may have made serviceable mine-cars and mobile homes, but the gods themselves did not connect fully with the culture of the wheel until the swift two-wheeled carts of Central Asia came into contact with the practice in Sumer of using wheels for royal or priestly display. It was then that the chariot's wheel, seen in profile, came to symbolize the orbs of the sun and the moon, and the noise of the thunderstorm came to be understood as the din of the gods' chariots streaking through the air.

Evidence of the fusion of Indo-European technology and Mesopotamian pomp and circumstance is not limited to the relief from Aleppo and the silver bowl from Afghanistan. To be sure, from the Egypt of Ramses II in the west to the China of the Shang dynasty (1600–1046 B.C.E.) in the east, the chariot symbolized royal majesty. But the dozens of Indo-European kings and warlords who ruled the lands in between outdid everyone else by conceiving of their entire warrior class as a separate order of society, divinely entitled to rule over the other orders of priests, craftsmen, and farmers.

The progenitors of today's Iranians and northern Indians labeled this order the *kshatriya*, and they enshrined their models of *kshatriya* behavior in the sacred Vedic texts, written in Sanskrit, which began to be composed around 1500 B.C.E. In one of these texts, mention is made of a festive event celebrating the intended union of Surya, the daughter of Savitr, a sun god, with the moon god Soma. An elaboration on that text composed in India around 600 B.C.E. describes the celestial gods competing in a chariot race run to celebrate the nuptials:

> By means of a mule chariot Agni [the fire deity] ran the race; as he drove on he burned their wombs; therefore they conceive not. With ruddy cows Ushas [the dawn deity] ran the race; therefore, when dawn has come, there is a ruddy glow; the form of Ushas. With a horse chariot Indra [the storm deity] ran the race; therefore its neighing aloud and resounding is the symbol of lordly power; for it is connected with Indra. With an ass chariot the Ashvins [twin sunrise and sunset deities]

THE RISE AND DEMISE OF THE CHARIOTEER

won, the Ashvins attained; in that the Ashvins won, the Ashvins attained, therefore is his [the ass's] speed outworn, his energy spent; he is here the least swift of all beasts of burden; but they did not take the strength of his seed; therefore has he virility and possesses a double seed.[9]

Since Central Asian chariot burials contain only horses, where did the mules and asses mentioned in the Vedic texts come from? This may sound like a trivial question, but it speaks to the fusion of the wheel tradition of Central Asia and that of Mesopotamia. The Mesopotamian practice of hitching onagers to light, two-wheeled carts arose well before anyone in Central Asia thought to harness horses to chariots. And even though visual depictions of domestic onagers do not survive after 2000 B.C.E., the practice must have migrated eastward to India, because the historian Herodotus, writing in the fifth century B.C.E., mentions Indian warriors in the army of the Persian king driving chariots drawn by "wild asses." Although he could conceivably be referring to one of the two half-ass species native to Central Asia and Tibet, the kulan and the kiang, there is no other evidence that either animal has ever been used domestically. His report, therefore, indicates an eastward diffusion of onager-drawn carts into Indo-European territory, for the Indians in the Persian army not only spoke an Indo-European language closely related to Old Persian, but also cultivated the memorization and study of the *Rig Veda*. Thus here is found another indication of chariot warfare involving the fusion of Central Asian technology and Mesopotamian cart-riding.

The surprise in the Vedic tale of the gods' chariot race is not just the unexpected mules and donkeys, but the victorious team of asses, driven by the divine Ashvin twins, whose name comes from the Vedic word for "horse." Asses can trot quite speedily, but they do not readily gallop. The author of the story explains away this fact by saying that the ass, which he knew as the "least swift of all beasts of burden," sacrificed his stamina to win, even though he retained the sexual potency that allows him to impregnate both a donkey and a horse. The reality behind the original myth, however, seems to be that onagers, not donkeys, drew the chariot

of the victorious Ashvins. This suggests that, at least for a brief time before the horse became inseparably identified with the chariot, the tradition of harnessing onagers to vehicles continued and contributed to the
creation of a chariot-centered warrior ideology. As for the mules that
pulled Agni's chariot, they were probably sterile crosses between onagers
and horses, not, as we now understand the term, between donkeys and
horses.

Just as the image of the Hittite storm god associates divinity with riding in a bull-drawn chariot, so the Vedic story of the chariot race suggests
that an equation between cart drivers and gods had taken form on a
Mesopotamian model before horse-drawn chariots came on the scene.
Hence the Indo-European chariot warriors who first drove into history
in Central Asia were able to add a new quality to their symbolic profile
once they entered the Middle East. They became not just warriors, but
avatars of the gods on high. And the gods continued to be represented
as chariot riders long after the cavalryman had superseded the charioteer as the dominant figure on the battlefield.

The symbolic turning point in the passage from godlike charioteer to
heroic horseman can be found in the legend of Alexander the Great and
the Gordian knot. Once upon a time in Phrygia, a kingdom that arose
before the Trojan War in the part of central Turkey that had once been
the homeland of the Hittites, an oxcart driver named Gordias lucked
into the job of king after an oracle proclaimed that the next driver to
enter the city, henceforth called Gordium, should rule. His son Midas
dedicated his father's oxcart to the supreme god and tied it in some
fashion with an exceedingly complex knot.

A millennium or so later, when Alexander the Great launched his
campaign against the Persian king Darius III, Phrygia had become a
province of the Persian Empire. But the Phrygians still revered the oxcart
of Gordias, with its intricate knot. Some versions of the legend maintain
that Alexander, becoming frustrated while trying to untie the knot, cut
through it with his sword; others, that he pulled out a pin and thus un-
coupled whatever it was that the knot was holding together. In any event,
by the time Alexander accomplished this feat in 333 B.C.E., legend had

transformed the oxcart into a chariot; and as time went on, the story expanded to include an oracular prediction that whoever loosened the Gordian knot would become the ruler of Asia. Hence, by performing that task, Alexander proved his destiny as conqueror of the Persian Empire.

In turning the oxcart into a chariot and making it an instrument of divine prophecy, the legend kept in step with changes in vehicular symbolism and technology. But an additional shift was in the offing. Alexander the Great was also the first historic hero to be known for his warhorse. At the age of thirteen, he subdued the powerful but previously untamable Bucephalos (Ox-head) and rode him throughout his conquests until the horse reached the end of his life in northern Pakistan. Alexander also made better use of cavalry during his campaigns than had any earlier ruler. Within a century of his confrontation with Darius III, a Greek artist captured the new dominance of cavalry over chariotry in a painting, now lost, that was copied two hundred years later on the mosaic floor of a building in Pompeii (figure 61). The Alexander Mosaic depicts the Macedonian conqueror on his mighty warhorse routing the Persian king, whose charioteer is whipping his horses into flight.

Thereafter, from the beginning of the Roman Empire in the first century C.E., the bronze statues of the emperors that ornamented the imperial capital showed them on horseback. Most of the statues were melted down to reuse the metal, but one that survives, depicting the emperor Marcus Aurelius (r. 161–180), typifies these representations of godlike warrior rulers (figure 62). Iran turned to the heroic symbolism of the mounted ruler at roughly the same time. In one bas relief, both the Zoroastrian god Ahura Mazda, who is holding a staff, and the Sasanid king, upon whom he is conferring power, sit astride mighty horses (figure 63).

In the Middle East, where the chariot had reigned supreme in 1200 B.C.E., the civilian economy as well as the military turned away from wheeled vehicles. Over the first six centuries C.E., not only chariots but also oxcarts largely vanished. Pack camels and donkeys, and the Arab

FIGURE 61 The Alexander Mosaic, from the House of the Faun in Pompeii and dating to 100 B.C.E, showing the mounted Alexander the Great (*far left*) defeating the chariot-borne Persian king Darius III, possibly at the Battle of Issus in 333 B.C.E. (Collection of the Museo Archeologico Nazionale, Naples)

tribes that bred them, provided transport services at low prices, not just along desert routes where the Arabs also levied tolls and provided caravan guides, but even in towns and villages. Over the same period, the Arabian horse emerged as the ideal riding animal. Thus the warriors who conquered an empire for the Islamic faith in the seventh century rode their camels to the battlefield but then mounted their warhorses when the fighting began.

In Europe, which lacked deserts conducive to camel breeding, oxcarts continued to carry loads for peasants, though cross-country traders more often used trains of pack animals as the Roman roads became dilapidated for want of repair from the third century C.E. onward. Even chariot racing, which had been added to the Olympic games in 680 B.C.E., passed into history after more than a thousand years of popularity. The

FIGURE 62 Equestrian statue of the Roman emperor Marcus Aurelius, second century C.E. (Collection of the Musei Capitolini, Rome)

last recorded race in Rome was run in 549 C.E., though chariots continued to race in Constantinople for an additional five centuries.

To be sure, vestiges of the tradition of divinely ordained monarchs traveling in wheeled vehicles survived for some time. But the derogatory phrasing of the ninth-century scholar Einhard's description of the

FIGURE 63 Relief at the necropolis of Naqsh-e Rustam, Iran, of the Sassanid ruler of Persia (*right*) receiving the symbol of power from the Zoroastrian god Ahura Mazda, third century C.E. (Photograph in author's collection)

conveyance used by the Merovingian kings who preceded Charlemagne, Einhard's horse-riding employer, on the throne of the Frankish realm (present-day France, Belgium, and western Germany) shows how weak the symbolism of the wheel had become: "When [the king] had to go abroad, he used to ride in a cart, drawn by a yoke of oxen driven, peasant-fashion, by a Ploughman; he rode in this way to the palace and to the general assembly of the people, that met once a year for the welfare of the kingdom, and he returned him in like manner."[10] Nevertheless, a few grand vehicles remained, as is apparent from the contents of the grave of a Frankish noblewoman excavated in central Germany and dated to the sixth century:

> The oldest and "richest" grave . . . is of a woman; it is placed at the centre and formed the origin of the burial site in question. It consists of a

huge wooden chamber measuring *c.* 3 by 5 metres [10 by 16 feet], and embedded almost 4.5 metres [15 feet] deep; the reconstruction shows a three-storey building. Here, in the first half of the 6th century, a woman was interred with a four-wheeled wagon and the harnesses for a team of horses which were buried in a nearby grave.[11]

Farther to the east, an account of the western Slavs (Czechs, Wends, and possibly Poles) written in Arabic around 965 states: "Their kings travel in big, rumbling, high vehicles with four wheels, and with four sturdy poles at the corners from which a brocade-covered *haudaj* [enclosed riding cubicle] is suspended by strong chains so that whoever sits in it does not feel the jolting. For the sick and the wounded they also used such conveyances."[12]

The vehicles of the Merovingian monarch, the Frankish noblewoman, and the Slavic king probably included Roman-era improvements in wagon design. The best Roman passenger vehicles had pivoting front axles for effective steering, even though the sharpness of their turns was constrained by the front wheels coming into contact with the wagon bed, a limit known as full lock. They also had a type of four-point suspension that allowed them to sway sideways on rough roads. A modern reconstruction of the best-attested Roman design (figure 64) looks very different from what modern French illustrators, following Einhard's disdainful reference to riding "peasant-fashion," have chosen to imagine for the Merovingian monarchs (figure 65). In the latter image, the true nobles are clearly the mounted knights in the foreground and not the *fainéant* (do-nothing) king lounging in the background.

What killed chariot racing and the tradition of royal carriages was not just the rise of armored knights as the dominating force on the battlefield, but also the inculcation of a tradition that members of the nobility who might aspire to knighthood—that is, males—should never demean themselves by traveling in a cart or wagon. In his poem *Lancelot, the Knight of the Cart*, the twelfth-century balladeer Chrétien de Troyes proclaimed this disdain in a tragedy of love. His hero, the Arthurian

FIGURE 64 A reconstruction of a Roman passenger wagon with a pivoting front axle and leather-strap suspension. (Collection of the Römisch-Germanisches Museum, Cologne; photograph by Marcuc Cyron / Wikipedia Commons)

FIGURE 65 *"Roi fainéant* (Royal Sluggard) of the Merovingian Race," a visualization of a sybaritic Merovingian king, who reclines in a wagon, compared with "real men," who ride horses. (From Edward Ollier, *Cassell's Illustrated Universal History* [London: Cassell, Petter, Galpin, 1890]. Private collection / © Look and Learn / Bridgeman Images)

knight of the round table, finds himself unhorsed during a frantic quest to rescue his kidnapped beloved, Queen Guinevere:

[T]he knight [found himself] all alone on foot, completely armed, with helmet laced, shield hanging from his neck, and with his sword girt on. He had overtaken a cart. In those days such a cart served the same purpose as does a pillory now; and in each good town where there are more than three thousand such carts nowadays, in those times there was only one, and this, like our pillories, had to do service for all those who commit murder or treason, and those who are guilty of any delinquency, and for thieves who have stolen others' property or have forcibly seized it on the roads. Whoever was convicted of any crime was placed upon a cart and dragged through all the streets, and he lost henceforth all his legal rights, and was never afterward heard, honoured, or welcomed in any court. The carts were so dreadful in those days that the saying was then first used: "When thou dost see and meet a cart, cross thyself and call upon God, that no evil may befall thee." The knight on foot, and without a lance, walked behind the cart, and saw a dwarf sitting on the shafts, who held, as a driver does, a long goad in his hand. Then he cries out: "Dwarf, for God's sake, tell me now if thou hast seen my lady, the Queen, pass by here." The miserable, low-born dwarf would not give him any news of her, but replied: "If thou wilt get up into the cart I am driving thou shalt hear to-morrow what has happened to the Queen." Then he kept on his way without giving further heed. The knight hesitated only for a couple of steps before getting in. Yet, it was unlucky for him that he shrank from the disgrace, and did not jump in at once; for he will later rue his delay. But common sense, which is inconsistent with love's dictates, bids him refrain from getting in, warning him and counseling him to do and undertake nothing for which he may reap shame and disgrace. Reason, which dares thus speak to him, reaches only his lips, but not his heart; but love is enclosed within his heart, bidding him and urging him to mount at once upon the cart. So he jumps in, since love will have it so, feeling no concern about the shame, since he is prompted by love's commands.[13]

The demeaned status and image of the wagon driver seemed as irreversible in 1200 C.E. as the rise of the charioteer as a paragon of royal and godlike power had seemed in 1200 B.C.E. In China, this change in status was never reversed. The chariots that had sped into battle in 600 B.C.E. had all but disappeared by 600 C.E., and the Chinese elite never again used wheeled transport to advertise their rank. In Europe, however, an amazing second reversal of image, a veritable "carriage revolution," took place between 1400 and 1650. Gender, which has played an important, if intermittent, role throughout the six millennia of wheel history, came to the fore in the carriage revolution—set in motion by men's insistence on women's limited and supervised travel, a restriction not imposed on the nomadic women of Central Asia.

seven

The Princess Ride

Boleráz women in the Carpathian Mountains may have sipped from clay mugs shaped like mine-cars, and Sumerian women surely witnessed the majestic processions of their monarchs and sacred statues. However, no specific evidence survives to connect women with wheeled transport in those early cultures. The same may be said of the practice of wagon nomadism that began on the Black Sea plain, but in that arena there is a stronger argument to be made from later sources for women playing not only an important role, but one that was to have a significant impact on the transportation history of Europe from the Renaissance onward.

Before discussing women and wheels in Europe, however, the subject of women and transportation at a more general level is in order. Until quite recently, men of high social status have regarded the mobility of women as a threat. (In Saudi Arabia, where women are not permitted to drive cars, they still do.) Ordinary women in foraging societies, nomadic tribes, farming villages, and poor urban neighborhoods sometimes enjoyed freedom of movement, but women of the elite classes almost always faced constraints. To the men who claimed responsibility for them—fathers, brothers, husbands—the unrestricted mobility of women carried with it the threat of dishonor. Like a precious piece of property, elite women had to be protected from covetous eyes. Nor could they travel without escort, since unworthy men might assault them or carry them off, to the disgrace of their male kin.

Restrictions on women's ability to walk with ease, if at all—such as bound feet, high heels, and skirts that are too full or too narrow—usually combine limited mobility with notions of sexual allure. Sexual concerns may also have been involved in moving high-status women from place to place if the society to which they belonged construed women's sexual appetites as being so aggressive that they would take advantage of any freedom from oversight to have sex with men. A'isha, the youngest wife of the Prophet Muhammad, for example, found herself accused of sexual misconduct after she went missing from a caravan returning to Medina. She showed up the next morning with a young man, who was leading her on his own camel. She explained that after she left her *haudaj*, a completely enclosed women's camel saddle, to answer the call of nature, and then tarried to look for a lost bangle, the men had loaded the empty *haudaj* back onto her camel without noticing that it was too light. The caravan departed, leaving her behind until the young man came along, perceived the situation, and, without gazing upon her, let her climb onto his camel to be led back to town. This incident, which was recalled over and over again during the following centuries as an example of the danger of allowing women freedom of movement, cast A'isha, though not the young man who helped her, in a sinful light. Tellingly, Muhammad asked A'isha's father, who had not been present during the incident, whether she was innocent before a Qur'anic revelation assured him that she was.

In the Islamic Middle East, a *haudaj* or a palanquin—in Europe, known as a sedan chair—were the only choices available for transporting high-status women (figure 66). Fully enclosed, a palanquin could be carried by either bearers or animals, though the latter were preferred.

> In Persia . . . the most dignified vehicle for traveling is the *takht ravan*.
> This is a large box with an arched roof, and a door-way at one end. . . . It

FIGURE 66 A Turkish palanquin (*takht-e ravan*) designed to be carried between two animals. (Photograph in author's collection)

is commonly five feet in length, nearly four in height, and about two and a half in breadth. . . . On each side there are staples, and by poles which are inserted into them, the vehicle is carried between two camels, mules, or horses. *This mode of conveyance is used chiefly by ladies of distinction.*[1]

In Central Asia, however, women not only dressed in trousers and rode astride horses, but also drove carts and wagons. Evidence for this comes primarily from medieval times rather than remote antiquity. However, it seems likely that women adopted these practices in the context of wagon nomadism as that style of life spread eastward across the Eurasian steppes from the Black Sea plain. Over the course of the second millennium B.C.E., wagon nomads speaking Indo-European languages trekked all the way to the frontiers of China.

At the eastern end of the Tarim River basin, China's northwestern province of Xinjiang, archaeologists have uncovered graves containing red-haired mummies wearing tartan plaids. Some of these people spoke and wrote an Indo-European language known as Tokharian, and some of them made use of wagons equipped with large wheels made of solid wood. Wagons with spoked wheels eventually replaced these cumbersome precursors just as an array of other peoples, speaking mostly Turkic and Mongolian languages, eventually replaced the Indo-Europeans during the first millennium C.E. However, all these peoples followed similar lifestyles featuring large herds of horses and fleets of wagons. When extensive eyewitness testimony became available in the thirteenth century, the wagons were described as being the responsibility of women.

The eyewitness report on Mongol transportation made by William of Rubruck, a Franciscan friar who journeyed from 1253 to 1255 to the court of the Great Khan Möngke, a grandson of the conqueror Genghis Khan, was one such piece of evidence. His remarks on women being in charge bear repeating:

> The married women make themselves very fine wagons. . . . When they unload the dwellings, the chief wife pitches her residence at the westernmost end, and the others follow according to rank. . . . Hence the court of one wealthy [Mongol commander] will have the appearance of a large town, though there will be very few males in it. . . . One woman will drive twenty or thirty wagons, since the terrain is level. The ox- or camel-wagons are lashed together in sequence, and the woman will sit at the front driving the ox, while all the rest follow at the same pace.[2]

Judging from sparse earlier reports, giving women the responsibility of moving the camp while the men were off herding horses or training for war dates back at least a thousand years before Genghis Khan to the time of the Indo-European Alans, among whom "all those who through age or sex are unfit for war remain close by the wagons and are occupied in light tasks."[3] To be sure, women are not explicitly mentioned as wagon drivers by the Roman soldier and historian Ammianus Marcellinus, who lived in the fourth century C.E. Moreover, Ammianus makes quite similar comments about the lifestyle of the Huns, a non-Indo-European steppe people who attacked the Alans and defeated them during his own lifetime. This suggests that Ammianus may be repeating a stereotype rather than reporting what he actually saw. But that from the time of the Scythians in the first millennium B.C.E., Greek and Roman writers typecast steppe nomads as wagon dwellers does not mean that the stereotype was unwarranted.

The question of who controlled the wagons in which the women rode distinguishes the nomads who traveled the eastern steppe from the Indo-Europeans who migrated into Europe. Both groups used wagons, but women in western and northern Europe did not control them. The root of this difference may be found in a change in horse herding among the Indo-Europeans who moved into the forests and mountains of Europe.

David Anthony, the scholar who has dealt with this issue most intensively, maintains that people probably began to ride horses in Kazakhstan before coming into contact with the Indo-European groups that were expanding into their lands from the west in wagons drawn by oxen. And even when techniques for harnessing horses to vehicles were developed, they were used mostly with chariots and other light carts. Oxen—or sometimes two-humped camels—were still being used to pull the heavier carts and wagons of the Mongols some four thousand years later. Hence the gendered division of labor intimated by Ammianus Marcellinus, and confirmed by William of Rubruck, represents a merging of two cultures: one based on setting up camps of ox-drawn wagons used as residences and the other founded on tending horses and riding them to war.

As we have seen, the cross-fertilization of these two technologies—that is, wheeled vehicles and horses—in the form of chariot warfare strongly affected the religious and social outlook of the peoples involved. The notion of a hereditary chariot-driving caste of warriors is deeply rooted in Indo-European myth and culture. But over time, the military obsolescence of chariotry caused a reversion to the underlying two cultures: ox-drawn carts and wagons for living, on the one hand, and horse riding for herd management, hunting, and war, on the other.

In summary, and recalling the discussion in chapter 4, the available evidence suggests that dwelling in wagons was a way of life that originated around 3000 B.C.E. among farmers who spoke Proto-Indo-European and inhabited the broad plain on the northern shore of the Black

Sea. Their vehicles had four solid wooden wheels and probably a light fabric roof, and they were drawn by oxen. As a drying climate between 3000 and 2000 B.C.E. turned forest to prairie, the spread of long grasses with deep, tenacious roots combined with limited precipitation to inhibit the continuation of grain farming. Some wagon dwellers migrated westward along river valleys and into northern Europe. Others trekked eastward into the valley of the Volga River and beyond, all the way to the frontier of China. For comparison, we might think of the wagon trains of pioneers opening the American West and of Afrikaner *Voortrekkers* moving into the northern regions of South Africa. As Ammianus Marcellinus said of both the Alans and the Huns, "in the wagons the males have intercourse with the women, and in the wagons their babes are born and reared; wagons form their permanent dwellings."[4]

East of the Volga River, the migrants encountered both a treeless steppe that made it necessary to conserve wood—hence the spoked wheels—and huge herds of wild horses that were already beginning to be domesticated by local nonagricultural peoples. The domestication of horses soon expanded in all directions, though oxen continued to pull the nomads' wagons. The use of spoked wheels also spread, and steppe artisans crafted some of the largest and most complex versions of them. At the same time, from Mongolia in the east to the Great Hungarian Plain, the westernmost extension of the Eurasian steppe into Europe, horse herding developed so extensively that horsemen began to live their everyday lives apart from their women, who assumed responsibility for the wagon camps. The largest assemblages of civilian wheeled transport anywhere in the world before the advent of the twentieth-century traffic jam were those organized by and presided over by Mongol women in the thirteenth century.

Given the role that women played in managing hundreds of carts and wagons and organizing their camps, it is not surprising that the Mongol and Turkic horsemen accorded them substantially more respect and imposed fewer forms of seclusion on them than did their Persian, Arab, and Chinese neighbors. Mongol women might even assume the reins of power over the entire empire when deliberations over succession followed the death of a ruling khan. According to an authoritative court historian, after Genghis Khan's son Ögödei passed away,

> Chaghatai [another son of Genghis] and the princes sent representatives to say that Töregene Khatun was the mother of the princes who had a right to the Khanate; therefore, until a *qurïltai* [family council] was held, it was she that should direct the affairs of the state. . . . Töregene Khatun was a very shrewd and capable woman . . . by means of finesse and cunning she obtained control of all the affairs of state and won over the hearts of her relatives. . . . And for the most part strangers and kindred, family and army inclined towards her, and submitted themselves obediently and gladly to her commands and prohibitions, and came under her sway.[5]

Women did not enjoy a similar degree of autonomy among the Indo-Europeans who migrated into Europe. They, too, rode in wagons, but they did not drive them. Judging from one surviving miniature painting, even the Cuman nomads, a steppe people who roamed the Great Hungarian Plain in the thirteenth century and spoke a Turkic language, lived and transported their women in vehicles driven and escorted by men on horseback (figure 67).

Southern Europe from Spain to Greece, where the influx of peoples from the steppe was much more limited, constituted a separate transport domain, one influenced more by speedy chariots than by lumbering wagons. The art of classical Greece presents a few depictions of women holding the reins of horses pulling chariots, but one cannot always discern whether the charioteer is a human or a goddess (figure 68). By Roman times, however, women drivers had largely disappeared, and two-wheeled carts predominated. A few Roman images of wagons hauling heavy cargo survive (figure 69), but most four-wheeled vehicles were used as conveyances for the social elite in general, rather than for women in particular (figure 70).

By the Middle Ages, most European images of four-wheeled wagons show women riding inside and men either driving the vehicles or

FIGURE 67 Cuman women in mobile homes being accompanied by men, thirteenth century. (Redrawn from the Radzivill Chronicle, fifteenth century)

FIGURE 68 A fresco of a woman, possibly the goddess Nike, driving a chariot, from Paestum, a Greek city on the coast of southern Italy. (Photograph by John McLinden / Flickr)

FIGURE 69 Relief of a Roman wagon carrying a heavy cargo.
(Gianni Dagli Orti / The Art Archive at Art Resource, NY)

FIGURE 70 Relief of a Roman passenger wagon. (Erich Lessing / Art Resource, NY)

controlling the animals by walking alongside them (figures 71 and 72). Thomas Moore's famous description of an imaginary society in *Utopia*, his novelistic fantasy published in 1516, supports this impression. Moore takes for it for granted that women need both an ox-drawn wagon and a nonelite male driver because real men travel on horseback:

> If any man has a mind to visit his friends that live in some other town, or desires to travel and see the rest of the country, he obtains leave very easily. . . . Such as travel carry with them a passport from the Prince, which both certifies the license that is granted for traveling, and limits the time of their return. *They are furnished with a wagon and a slave, who drives the oxen and looks after them; but, unless there are women in the company, the wagon is sent back at the end of the journey as a needless encumbrance.*[6]

The alternative of women riding on horseback, which seems obvious to our modern experience, had two drawbacks. First, skirts interfered

FIGURE 71 A miniature of royal women being conveyed in a wagon, accompanied by a postilion and other men, illustrating psalms 102 and 103 in the Luttrell Psalter (early fourteenth century).

THE PRINCESS RIDE

FIGURE 72 An illumination of women riding in a wagon, accompanied by a postilion and other men, in the *Weltchronik* (thirteenth century) of Rudolf von Ems, a Swabian (Austrian) epic poet.

with sitting astride a mount, and the posture itself was considered indecent, as was the wearing of trousers. Second, sitting sideways on a horse provided a very precarious perch until the seventeenth century, when secure sidesaddle designs were developed. Consequently, even so august a personage as Queen Elizabeth I "often used to ride [a horse], on state occasions, on a pillion [cushion behind the saddle], [sitting on the same horse] behind the lord chancellor or the lord chamberlain."[7]

Where discreet and secure vehicular means of transporting elite women from place to place were unavailable, as they were in most parts of the world, leaving women at home seemed to be the best policy. Although the seclusion of high-status women in the lands of Islam developed from a combination of many cultural factors, the absence of suitable transportation reinforced it. The situation was little better in other non-European lands. The two-wheeled carts used in India, China, and Southeast Asia afforded limited privacy—most had no roofs or sides—and scant space for handmaidens or chaperones. And wherever aristocratic men risked disgrace by being seen driving or riding in a cart, as was the fate of poor Sir Lancelot in the poem by Chrétien de Troyes, a woman of high status would have had to sit next to or behind a driver, who might be a servant or a slave. This clearly would have been demeaning for a proper lady.

Palanquins, of course, made it possible to move women about in total privacy when it was absolutely necessary, and palanquins carried on the shoulders of bearers became popular in China and Japan. Since a palanquin held at most two passengers, however, this mode of transportation deprived elite women of their female attendants on trips of significant length. Accordingly, they used palanquins most commonly for short journeys.

Palanquins also found favor with the gentlewomen of Europe. In addition to privacy, they kept their passengers out of the mud and manure so common on city streets. Yet as figures 71 and 72 and the passage from Moore's *Utopia* make clear, European societies never totally abandoned four-wheeled passenger wagons, which provided a means of transporting parties of elite women over long distances. For a thousand years, from the end of the Western Roman Empire to around 1500, pictures of European passenger wagons, usually covered for privacy, almost always show the passengers to be female. And on the outskirts of Constantinople (present-day Istanbul), the cosmopolitan center of the Eastern Roman Empire, parties of women were still picnicking in special wagons drawn by oxen and conducted by men well into the nineteenth century (figure 73).

FIGURE 73 Women riding in a picnic wagon outside Istanbul, nineteenth century. (Photograph in author's collection)

Moving a lady and her attendants by wagon, instead of by palanquin or on a horse, with each woman sitting sideways behind a horseman, constituted a singular feature of European society. Taking into account the aversion felt by aristocratic men in the age of knighthood to riding in two-wheeled vehicles because of their association with crude farming folk—the "miserable, low-born dwarf" cart driver in *Lancelot, the Knight of the Cart*—it is apparent that the four-wheeled wagons used by aristocrats served a different purpose than the two-wheeled carts used by commoners. Like their counterparts in China and India, both the rural folk of the countryside and the commercial haulers of bulky goods recognized that carts were more efficient than wagons because they were more maneuverable and the cart's one axle generated less friction than the wagon's two (figure 74). But the tradition of carrying dignified passengers in four-wheeled wagons, which dates back to the wagon burials

FIGURE 74 Jesus sows seed, which then grows and the grain is reaped and carried in a medieval English farm cart with an extended bed, in an illustration in MS Selden Supra 38 (early fourteenth century). Compare this working cart with the passenger wagon shown in the Luttrell Psalter (figure 71), which dates to the same period.

on the Black Sea plain in the fourth millennium B.C.E. and was bolstered by technical improvements in steering and suspension made during the Roman Empire, persisted through the centuries despite the economic disadvantages of wagons vis-à-vis carts.

The difference between women driving and being in charge of wheeled transport in Mongolia, at the eastern end of the Eurasian steppe, and riding docilely in wagons controlled by men at the western end deserves emphasis. Both practices originated with the diffusion of Indo-European wagon nomads from the Black Sea plain, and elite Mongol women in wagons enjoyed greater mobility and less seclusion than their sisters to the south and east, whom custom and religion either confined to their homes or limited to short journeys by palanquin. But simply by preserving the notion of using wagons for elite personal transport, the male-supervised mobility of women in eastern Europe set the stage for the "carriage revolution," starting with the invention of the coach.

eight

The Carriage Revolution

I n 1400, European aristocrats were split along gender lines on the matter of riding in wagons. Unless they were old and infirm, men rode horses and shunned wheels. When noblewomen undertook a journey, however, they and their attendants traveled by wagon. This divide came to an end over the course of the fifteenth and sixteenth centuries. This chapter proposes that advances in military technology in the central European theater of war, featuring Hussites against Catholics, on the one hand, and Ottoman Muslims versus European Christians, on the other, triggered a change in men's thinking that led them to believe that coaches and carriages could suit the status of a knight just as well as a fine warhorse and polished armor. This period also marks the height of the Renaissance. Appropriately, therefore, Renaissance palaces like the Hôtel de Soissons, built in 1610, included for the first time a porte cochere, an arched entryway through which carriages could enter, drop off passengers, and then leave.

Students of European transportation during the medieval period agree that women rode in wagons and men did not: "Abundant records of the English royal household and aristocratic families allow a continuous picture from the thirteenth to the sixteenth century of the status of horse and carriage and their use, predominantly by women, in royal courts and noble households."[1] This was common knowledge in the nineteenth century, when the history of the carriage was a topic of general interest in Europe and the United States and the rise of the automobile had not yet lured readers in another direction. Witness this from 1871:

> Carriages that can in any way be considered as the precursors of our present elegant vehicles were not invented or brought into use until the Middle Ages; and for many years after the first introduction of coaches they were only used by ladies, and by them only upon state occasions. Kings and knights considered all kinds of carriages as effeminate machines, and scorned to be seen within them . . . As late as the reign of Francis I [r. 1515–1547], there were only three coaches in Paris. One of these belonged to the Queen, another to Diana of Poitiers [a favorite of Henry II, Francis's successor], and the third to René de Laval, a corpulent nobleman, who was unable to ride on horseback.[2]

None of these histories, however, offered much explanation of why the sixteenth century marked a turning point, by the end of which carriages were becoming increasingly popular with men of aristocratic station. Some writers opined that new technologies made sixteenth-century vehicles more functional and comfortable. Specifically, they referred to the "rediscovery" of the pivoting front axle, which had facilitated the steering of wagons in the Roman era, and to the "new" idea of suspending the passenger compartment on transverse or longitudinal chains or leather straps, which allowed it to swing and sway instead of harshly jolting over almost universally terrible roads. More recent scholars, however, have challenged both these claims and mustered evidence to show that improved steering and suspension technologies were already known before 1500.[3]

These technical debates cannot be resolved definitively in a way that explains the sixteenth century as a watershed in Europe's transport history. The few vehicles and images of vehicles that survive from before 1500 do not reveal a technological uniformity among carts and wagons

FIGURE 75 The bridal carriage of Princess Dorothea of Denmark, with a rudimentary leather-strap suspension, mid-sixteenth century. (Collection of the Kunstsammlungen der Veste Coburg, Coburg, Germany)

in different parts of Europe. Wheelwrights—whose assembling of spokes, felloes, hubs, and rims called for much skill and experience—formed an identifiable trade, but carriage makers and wagon builders did not. The farmers and haulers who built and drove most of the wheeled vehicles in use drew on technical traditions that varied widely from one region to another, and even within a region, as evidenced by the coexistence in Turkey and Portugal of oxcarts with wheelsets and others with independently rotating wheels. As for the much smaller number of vehicles servicing the needs of royalty and noblewomen, their decor commanded more attention than their technical design (figure 75). Noble or royal weddings inspired the most lavish vehicles. The nuptial journey of Beatrice of Naples to wed Matthias Corvinus epitomizes the importance of style:

The Princess, then bride-elect to Matthias Corvinus, King of Hungary, came to Buda in 1476 with a large suite. The bridegroom-King, accompanied by three thousand mounted noblemen, went to meet her at Fehérvár where she arrived in a gilded coach covered with gold embroidered green velvet. The members of the Princess' suite traveled in seven richly gilded coaches, each of which was drawn by six horses; the coachmen wore velvet suits with gold buttons. Matthias brought with him two coaches from Buda, which were even more magnificent than those of his bride-elect; after the Fehérvár festivities they entered Buda in the Hungarian vehicles.[4]

By 1600, however, riding in carriages was taking off all over Europe, with noblemen as likely to avail themselves of these vehicles as their wives, mothers, and sisters (figure 76). Improved comfort cannot explain the new rage for traveling by carriage, for deeply rutted and muddy roads,

FIGURE 76 A papal carriage with elaborate suspension, nineteenth century. (Library of Congress, Prints and Photographs Division / George Grantham Bain Collection)

FIGURE 77 An overturned Hungarian carriage, 1483. Note the absence of a pivot in the middle of the front axle. (Photograph in author's collection)

not to mention banditry, made cross-country travel as arduous in 1600 as it had been in 1400—so arduous, indeed, that coaches venturing between cities sometimes carried bundles of straw to soak up mud, poles and rope for pulling the conveyances out of ditches, and extra coachmen to help deal with such predictable calamities (figure 77).

Nor did carriages become popular because of government encouragement. Numerous kings and princes inveighed against what one decried as *effemination asiatique.* In 1588, for example, the German prince Julius von Braunschweig prohibited pleasure driving in fear that "the manly virtues, dignity, courage, honour and loyalty of the German nation were impaired, as carriage-driving was equal to idling and indolence"; and the contemporary satirist Johann Fischart asserted that riding in coaches was superseding riding on horses and thus was depressing the German saddle-horse industry.[5] László Tarr, the Hungarian carriage historian who cites these examples, says of his own homeland:

> In Hungary carriage driving was, one might say, a veritable national custom; in the sixteenth century it was so much in fashion even among the lesser nobility that, for love of comfort, many a nobleman went to war in a coach instead of on horseback or on foot. In view of these circumstances it had to be decreed by law that officers were to join the army on horseback or on foot, as befitting camping soldiers and *non in kocsi* [not by coach] as was their wont.[6]

Kocsi, the word specified by this decree, points to Hungary as the land where men of rank first switched from horses to carriages. Kocs (pronounced "coach") was a stopping point on the road that ascended the valley of the Danube River from Buda, the capital of Hungary, to Vienna, the de facto capital of the Holy Roman Empire. For reasons that are far from clear, the name of this village became attached to the four-wheeled vehicle, called in Hungarian a *kocsi* (pronounced "coachee"), that inspired, or represented, the new mode of travel throughout central and western Europe. The Slavic peoples immediately to the north and south of Hungary borrowed the word (for example, Czech *kocár* and Serbian *kochiye*), but those farther to the north and east—such as the Poles, Russians, and Ukrainians—did not, nor did the neighboring Romanians, though the flat, treeless Great Hungarian Plain, which probably contributed to the new vehicle's popularity, extends into the western part of Romania. The word "coach," which in English eventually became interchangeable with "carriage," traveled mostly westward: German *Kutsche*, Dutch *koets*, French and Spanish *coche*, and, perhaps earliest, the mention in an Italian document of 1487 of a vehicle variously called *careta de Kozo*, *caro de Coki*, or simply *Cozy*.[7]

The once-common view that the Hungarian *kocsi*, beyond being light and fast, embodied some specific technological advance in either steering or suspension has faded upon the examination of specific vehicles called coaches from various parts of Europe and the failure to find any uniform or significant technical improvements. Moreover, the earliest representation of a genuine Hungarian *kocsi*, published in 1568, a century after its putative invention during the reign of Matthias Corvinus (r. 1458–1490), lacks the suspension devices found on coaches elsewhere, just as coaches elsewhere do not have the wicker body of the Hungarian specimen (figure 78).

The riddle, then, is why Europe's aristocracy should have been so drawn to some sort of Hungarian passenger wagon in the sixteenth

FIGURE 78 A reconstruction of a Hungarian *kocsi* from the sixteenth century. (Photograph in author's collection)

century that the word for it entered their several languages. Julian Munby, a specialist on medieval European travel, glosses over the problem:

> The appearance of the coach is an extraordinary story that bears little relation to technological change, but has not been fully explained before, though the outline of the story is reasonably well understood. The *kotchi wagen* originated in Hungary at the court of King Matthias Corvinus . . . as a fast and light vehicle for men (rather than women), and was named after the small posting town of Kocs on the road between Budapest and Vienna. Their reputation spread from Hungary around the empire, and the queen's nephew, Hippolito D'Este, took one back to Italy in c. 1500 with its Hungarian driver. After a few decades of gradual expansion of use in Italy and elsewhere there was then a sudden and popular introduction of the "coach" around 1550 in all the capitals of Europe.[8]

The "reasonably well understood" story that Munby lays out suggests a well-attested pattern of innovation diffusion following a logistic curve, a statistical measure related to the bell-shaped, or "normal," distribution curve commonly used in grading. It is shaped like a stretched-out *S*, starting low, growing rapidly in the middle, and tapering off slowly at the top. The adoption curve of a typical "new idea" begins with one or more innovators. It expands slowly through the enterprise of a small but growing number of early adopters, who are usually people from social, economic, and educational backgrounds that make other people take notice of what they do. Then the new idea takes off when the example set by the early adopters catches fire. Finally, the curve trails off when everyone who is anyone has had a chance to try the idea.[9] The curve suggested in figure 79 does not imply that every European nobleman had purchased a coach by, say, 1650. It simply shows the chronological pattern by which the aristocracy learned about, and became enamored of, something new in the world of wheeled transport.

Specialists on innovation diffusion have validated this pattern of adoption in thousands of cases, but in most examples, such as the spread of automobile driving or bicycle riding, specific technological

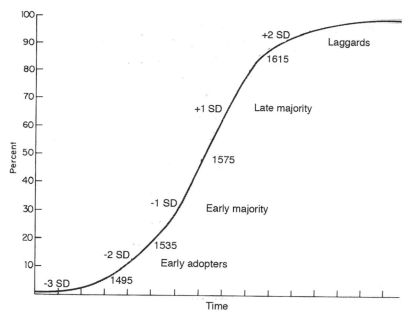

FIGURE 79 A hypothetical logistic curve of the adoption of the use of coaches in Europe, 1450–1650. SD refers to standard deviation from the mean.

innovations can be traced with quantifiable data. As Munby points out, however, although the coach belongs to the realm of technology, it was not a game-changing innovation like radio or television. In fact, its appearance "bears little relation to technological change." What altered instead, in epidemic fashion, was the attitude of elite European men toward riding in vehicles. The question to ask, therefore, is what changed in their worldview between roughly 1450 and 1650? And what did Hungary have to do with it?

Most European noblemen imagined their life trajectories to involve, at some point, going to war, and consequently they trained from childhood in riding and fencing. According to the biographer of an eminent twelfth-century knight:

The horse was a primary marker of knightly status, but also an intimate companion in combat, an animal in whom the rider placed his trust, perhaps even his life. Not surprisingly, favored mounts were pampered, treasured, sometimes even named. . . . The most important [weapon] was the sword—the totemic weapon of knighthood. It played a central role in the ritual of knighting, and both the carrying of a sword, and the ability to show skill in its use, came to be intimately associated with this elite warrior class.[10]

They thought about warfare and followed accounts of wars currently being fought. Throughout Europe, armored knights on horseback dominated the battlefield beginning in the age of Charlemagne (r. 768–814), and they dominated the aristocratic imagination as well. Indeed, knightly combat emerged as the defining characteristic of European nobility, and a knight's colors and heraldic symbols adorned his military equipage, including his horse.

The spread of gunpowder weapons capable of stopping a charging knight eventually brought this era to an end, but the power and accuracy of guns developed slowly, and heavy body armor could still stop bullets well into the seventeenth century. In eastern Europe, however, the light cavalry tactics of Mongol and Turkish steppe nomads had caused heavy armor to fall out of use a century or so earlier. Yet even without armor, officers almost always rode horses and continued to do so right through the period of the American Civil War.

From the fifteenth century until well into the seventeenth, Christian monarchs and clergy reckoned that the Muslim Ottoman Empire was Christendom's foremost enemy. Although the first crude cannons came into use in Europe in the fourteenth century and were commonly employed in sieges beginning in the fifteenth, the artillery used in the Ottomans' conquest of Constantinople in 1453 stood out in terms of size and effectiveness. One gun built for the sultan is said to have been dragged 150 miles to the city's walls by two hundred men and sixty oxen. Notably, the cannon founder who forged this behemoth was a Hungarian named Orban.

Guns that could be carried and fired by a single soldier were too heavy and cumbersome to be wielded by a man on horseback. Nor could they be aimed accurately. Thus gunpowder was apportioned between cannons installed in fortresses or aboard ships and heavy hand weapons, called arquebuses, fired by infantry. The earliest and most effective infantry to be trained in these arms were the janissaries of the Ottoman Empire, but Matthias Corvinus employed a large number of arquebusiers—infantry gunmen—in his famed Black Army in the late fifteenth century. From there, the practice spread, and in 1503, at the Battle of Cerignola, arquebus fire gave Spain the victory over a French army that did not have arquebusiers.

The innovation of mounting a cannon on a pair of wheels that could be drawn by oxen is credited to Jan Žižka, the brilliant, and for the last three years of his career totally blind, Czech general who commanded the followers of Jan Hus, a proto-Protestant dissenter who was burned at the stake in 1415. The ensuing Hussite wars against Catholic monarchs wracked Bohemia (present-day Czech Republic) between 1418 and 1424.

Bohemia was also the one part of Europe where the development of gunpowder weapons overlapped that of four-wheeled vehicles. Possibly inspired by the tradition of the Huns, Cumans, or other wagon nomads who from time to time had occupied the Great Hungarian Plain—the fourth-century Roman strategist Vegetius remarked that "all the barbarous nations range their carriages round them in a circle"[11]— Žižka chained the Hussite wagons together to form a circle and placed fourteen to twenty soldiers and two drivers in each wagon. Two of the soldiers fired small cannons, and the others wielded crossbows, pikes, or flails to keep the enemy at bay while the gunners reloaded. The small cannons firing through the gun ports of the wagons, and the larger ox-drawn cannons placed between them, would wreak such havoc that enemies would be goaded into a futile attack and thus open themselves up to a counterattack by Hussite horsemen from inside the wagon circle (figure 80).

Žižka never lost a battle, and his wagon-circling tactic, known as a laager or, in German, a *Wagenburg* (wagon town), deeply influenced

FIGURE 80 Hussite warriors fighting from within a circle of linked wagons, fifteenth century. Note the arquebusier firing to the left and the larger cannon tubes in the wagon at the top. (Wikipedia Commons)

military thinking in Europe's embattled eastern war zone. In 1444, for example, the Ottomans inflicted a crushing defeat on the flower of east European knighthood at the Battle of Varna in eastern Bulgaria. The representative of the pope, Cardinal Julian Cesarini, who had earlier failed in a Catholic crusade against the Hussites, urged his fellow leaders to take refuge within the Hussite-style wagon circle organized by the commander of the Hungarian contingent; but they opted for the unwise choice of charging uphill into the Ottoman lines. Cesarini died; the Hungarian commander lived.

Thirteen years later, the Hungarian Diet chose as king that commander's son, Matthias Corvinus, who ruled as one of the most erudite and cultured monarchs of the Renaissance. Corvinus fought for ten years, from 1468 to 1478, to win control of Bohemia and enlisted many Hussite army veterans in his mercenary Black Army. One-quarter of these soldiers were armed with arquebuses, and they continued in some battles to follow the Hussite tactic of firing from the protection of circled wagons. The proportion of gunpowder soldiers in the Black Army was almost triple that in the armies of western Europe.

Did Hussite military tactics spread by way of Hungary and coincidentally spark the runaway popularity of carriage riding that became manifest a century and more later? So I will argue. Just as in the United States, male attitudes in late medieval Europe tracked the changing theaters and styles of warfare. The crusades of the twelfth century gave rise to a holy war vocabulary, improvements in castle design, and the foundation of religious orders of knights that echoed through every part of Europe. The thirteenth century saw a similar change in the military imagination. The conquests of Genghis Khan replaced the imperial Roman specter of "barbarians at the gate" with the terror of bloodthirsty Mongol horsemen. Another century passed, and the Hundred Years' War between France and England, ending in 1453, and the Christian *reconquista* of Spain from the Muslims, continuing to 1492, together focused the male imagination on gunpowder versus armor and longbows versus crossbows.

These fifteenth-century wars in western European coincided with a fresh military threat to Christian Europe in the form of the Ottoman Empire. An Ottoman sultan destroyed the Serbian kingdom at the Battle of Kosovo in 1389, and another cut down many of Christendom's boldest knights at the disastrous Crusade of Nicopolis in Bulgaria seven years later. Thereafter, eastern and central Europe became the theater of war par excellence, and the military prominence of the region continued with the sixteenth-century wars of religion between Protestants and Catholics, of which the Hussite wars had provided a foretaste. To be sure, the exploits of European navies in the Indian Ocean and the triumphs of Spanish conquistadors against foes with neither gunpowder nor armor in the New World held a strong grip on some imaginations, but practical military minds concentrated on the battlefield role of cannon and handguns, and what that role might imply for the next phase in the history of the mounted knight.

In evidence of this, witness the spread westward of two words, "howitzer" and "pistol," that derive from the same military context as the Hussites' wheel-borne artillery. The word "howitzer," referring to a type of short-barreled cannon that used comparatively little gunpowder to lob projectiles in long parabolic arcs, comes from the Czech word *houfnice*, designating a type of artillery first deployed by the Hussites. It shows up in German as *Aufeniz* in 1440, metamorphoses into *Haubitze*, and then spreads in various pronunciations into Swedish, Finnish, Polish, Russian, Italian, Spanish, Portuguese, French, Dutch, and English. As for the word "pistol," meaning a small handgun, it is thought to derive from the Czech word *pišt'ala*, designating a type of Hussite hand weapon, and to have entered French and English in the mid-sixteenth century by way of the German *Pitschale*, *Pitschole*, and *Petsole*.

What these language trends indicate is that the tactics and armaments that emerged in the central European military theater, starting with the Hussite wars in the first quarter of the fifteenth century, increasingly captured the imagination of the European aristocracy as stories about them filtered westward. From the first Ottoman siege of Vienna in 1529 onward, central Europe became the paramount arena for European land wars. This was where new types of gunpowder weapons were deployed and where the myth of the heavily armored knight as lord of the battlefield would be shattered.

The Hussites' notorious use of circled wagons as gun platforms, which was imitated not just by Christian warriors but also by Muslim military men, provided for Europe a hitherto absent vision of wheeled vehicles that were also manly, as opposed to the ladies' wagons and ornamented bridal conveyances that for centuries had defined four-wheeled passenger transport for the elite. Not only did Matthias Corvinus, who enjoyed wide renown as a paragon of both military prowess and Renaissance sophistication, endorse the coach as a fast, lightweight men's vehicle, but the first evidence of its spread westward, into Italy, is linked directly to his court. If the king and his heroic Hungarians could ride in coaches, then why couldn't noblemen elsewhere follow his example? Why couldn't carriages emblazoned with heraldic devices and manned by crews of drivers and footmen wearing their masters' colors take the place of knightly shields as billboards for aristocratic display?

Throughout this book, I have argued that new or improved wheel technologies arise when changing circumstances create new needs and thus open up opportunities for innovation. In this instance, the opportunities were military, but the broader impact was psychological. In summary, here are the broad strokes of my argument:

1. The tradition of families traveling in four-wheeled wagons spread into northern Europe from the Black Sea plain beginning in the third millennium B.C.E.

2. After 2000 B.C.E., two-wheeled vehicles supplanted four-wheeled wagons for most purposes. War chariots dominated the battlefield, but then became obsolete and fell out of use by the beginning of the common era. Two-wheeled carts prevailed in the peasant economy, but four-wheeled vehicles still found occasional use in the Roman Empire, mainly as elite passenger vehicles.

3. As the armored knight mounted on his charger increasingly came to define the European male aristocracy from 800 C.E. onward, passenger wagons began to be used almost exclusively for transporting women of rank with their female entourages.

4. During the sixteenth century, European noblemen gradually abandoned their disdain for wheeled transport and took up riding in coaches and carriages.

5. This change in the attitude of aristocratic men toward wheels proceeded westward from central Europe and paralleled the emergence of that region as the continent's paramount theater of war.

6. The westward spread and adoption into other European languages of the Hungarian word *kocsi* and of Czech words for gunpowder weapons point to the Hussite wars in the first third of the fifteenth century as the seedbed for this far-reaching change in the mentality of elite men.

By the early seventeenth century, the bandwagon phase of the carriage revolution was well under way in England. In 1555, a chronicler had remarked: "This year Walter Rippon made a coche for the Earl of Rutland, which was the first coche that ever was made in England." A generation later, in 1601, a bill was debated in Parliament "to restrain the excessive use of coaches," and in 1623, a Thames boatman, angered by wheeled competition, wrote of "needless, upstart, fantastical and time-troubling" coaches "damming up the streets and lanes."[12] The surge in the popularity of carriages generated new problems and exacerbated old ones: traffic, manure, and mud on the streets; broken paving stones and deep ruts caused by iron-rimmed wheels; and the social elite owning multiple vehicles. The countryside, where roads remained rough and poorly maintained, was slower to be affected. On the more positive side, carriage building became a prestigious occupation, the demand for skilled drivers and footmen rose, and riding a horse ceased to connote knightly violence.

Some two centuries later, the carriage revolution reached a totally unexpected climax with the idea of mounting a steam engine on wheels. Only in Europe, and in a few of European nations' overseas colonies, did four-wheeled mine-cars and carriages, along with the iron rails and paved roads that evolved with them, provide the infrastructure that was needed to make the transition from horses and oxen to steam power and gasoline engines.

The world's other wheel-using societies continued to satisfy their transport needs with two-wheeled carts. Lest one think that this was a sign of backwardness, however, one should remember that carts were easier to build than wagons, operated with less friction, and did not require a complicated steering apparatus. Indeed, it can reasonably be argued that the carriage revolution, which, as has been shown, was not triggered by industrialization or "modernity," not only set Europe apart from the rest of the world, but also reflected a penchant for complicated technologies—steering, suspension, brakes—that did not necessarily add up to superior economic efficiency.

Technologically, there was nothing to prevent China or India from developing a similar four-wheel infrastructure. Both regions bordered the Eurasian steppes, home to the wagon nomads; both regions adopted chariot warfare from them and then followed them in abandoning chariots in favor of mounted cavalry. What set China and India apart from Europe in not developing carriages was not the absence of skills, materials, or inventiveness. But, similarly, it was not a matter, as Jared Diamond has contended, of chance-driven geographic proximity to draft animals and east–west corridors that connected East and South Asia with the Middle East and Europe leading ineluctably to the rise of modern Europe. Wheeled land transport was a key element in this rise, but the carriage revolution that set the stage for Europe's unique development of motor transport was psychological—how men and women thought or felt about wheeled transport—rather than technical or economic. In China, though, the psychological impetus and imperial endorsement for the adoption of four-wheeled vehicles was absent.

nine

Four Wheels in China

"If you build a better mousetrap, the world will beat a path to your door." This old saying underlies the dismay so often expressed at the absence of wheeled transportation in the pre-Columbian societies of the New World. Wheeled toys provided the model for a better mousetrap, at least in the minds of the puzzled modern commentators, but no one beat a path to the toymaker's door. Although no one has ever mentioned it, Chinese history shows a similar gap between hypothetical promise and practical fulfillment. The concept of the four-wheeled vehicle that in the form of the carriage transformed Europe's transportation economy beginning in the seventeenth century was known in China, but no one seems to have thought that it represented a better mousetrap. So China had no carriage revolution. Modern observers brought up in societies that revere wheels in general, and four wheels in particular, should understand that for inventions to spread and be successful, they must address an existing need, even if it is a psychological one.

Before the Mongols conquered China and established the Yuan dynasty (1271–1368), China had no four-wheeled vehicles at all. Being direct inheritors of the wagon-nomad tradition of the Eurasian steppes, the Mongols introduced four-wheeled transport into China. And the innovation stuck. It can be traced to at least the mid-eighteenth century, four hundred years after the Mongols had fallen from power. But its impact was slight and did not include carriages. Thus China did not develop the steering, braking, and suspension technologies that marked the carriage era in Europe or the European concern with improving road surfaces.

We have already discussed William of Rubruck's mid-thirteenth-century account of his travels among the Mongols and highlighted the dominant role that women played in managing the hundreds of wagon-mounted dwellings and baggage carts that made up an encampment. The friar goes on, however, to specify the scale of some of the vehicles:

> And they make these houses so large that they are sometimes thirty feet in width. I myself once measured the width between the wheeltracks of a cart [at] twenty feet, and when the house was on the cart it projected beyond the wheels on either side five feet at least. I have myself counted to one cart twenty-two oxen drawing one house, eleven abreast across the width of the cart, and the other eleven before them. The axle of the cart was as large as the mast of a ship, and one man stood in the entry of the house on the cart driving the oxen.[1]

What did this mobile home look like? How many wheels did it have? How were the animals harnessed for pulling it? A width of thirty feet would span two and a half lanes of an American interstate highway—WIDE LOAD indeed. The monster vehicle that William of Rubruck measured would all but fill a modern neighborhood street. As for the two rows of eleven oxen pulling shoulder-to-shoulder and haunch-to-haunch, one can only wonder how they were controlled. A yoke resting on the neck and pulling against the upward protrusion of the bovine thoracic vertebrae accommodates only two oxen. When more oxen are needed, the practice in Europe and America has always been for one yoked pair to be harnessed in front of another (figure 81).

FIGURE 81 Multiple yoke of oxen pulling a heavy wagon in South Africa, late nineteenth century. (Istock / duncan1890)

It comes as no surprise, therefore, that a sixteenth-century engraver charged with illustrating William of Rubruck's narrative came up with a *yurt* on four wheels being pulled by a long line of eleven yokes (figure 82). This, despite William's explicit statement that the two rows of animals were harnessed abreast across the front of the vehicle. A nineteenth-century illustration of the same passage sticks closer to the text, though it also visualizes four wheels, while William mentions only one axle (figure 83). But the artist was puzzled by the harnessing. Instead of yokes, he shows a strap across each animal's breast. Breast straps were commonly used for harnessing horses in both Europe and China, but the technique is ill-suited to oxen, which hold their heads too low to make the breast a plausible point of traction. The artist also drew the front wheels with a smaller diameter than the rear wheels, a typical indicator of a pivoting front axle. But a vehicle with a twenty-foot axle

FIGURE 82 A Mongol mobile home being pulled by twenty-two oxen, as shown in a sixteenth-century illustration of William of Rubruck's description of encampments in his mid-thirteenth century account of his travels.

FIGURE 83 A Mongol mobile home being pulled by twenty-two oxen, as shown in a nineteenth-century illustration. (From *The Travels of Marco Polo*, trans. Henry Yule, ed. Henri Cordier [London: Murray, 1920])

as thick as a mast could not plausibly be steered by that means, and a *U*-shaped bracket on the outside of the front wheel makes it clear that the illustrator did not imagine the vehicle turning. A still more recent effort to envision what William of Rubruck saw places an enormous four-wheeled vehicle in a military rather than a domestic context and suggests that somehow all the oxen pulled with breast straps on a single wagon tongue extending from the middle of the front axle (figure 84).

These fantastical imaginings might tempt William's readers to suspect that despite his reputation for accuracy, he indulged in gross exaggeration when it came to describing these giant mobile homes. Fortunately for his reputation, evidence from China largely confirms his report. In 1736, the Qianlong Emperor (r. 1735–1796) of the Qing dynasty (1644–1912) was presented with a meticulously realistic scroll painting. Thirty-eight feet long, it depicts pedestrian and vehicular traffic along a river flowing through a Chinese city and the surrounding countryside. The five artists

FIGURE 84 A Mongol mobile home, as shown in a twentieth-century illustration.

FIGURE 85 Mongol-style harnessing of a four-wheeled wagon hauling a building stone, as depicted in the eighteenth-century version of Zhang Zeduan's scroll painting *Along the River During the Qingming Festival* (twelfth century). (Collection of the National Palace Museum, Taipei, Taiwan)

who collaborated on the painting detailed thousands of individuals and some eighty conveyances. Among the latter was a four-wheeled wagon drawn by twenty-two horses or mules arranged in two lines, one in front of the other (figure 85). Each animal is harnessed by means of ropes tied to either side of a horse collar, and the ropes are then gathered behind them to form two thick cables that reach to the front of the wagon. Judging from the height of the twenty or so men attending it, the wagon's spoked wheels are about four feet in diameter and of equal size, which means there was no provision for turning corners. Using the same size comparison, the stone that the wagon is transporting seems to be about four by four by six feet. If, as is likely, it is a granite, marble, or limestone building block, it should weigh around eight tons. This amounts to a fairly realistic 750 pounds per animal. As for the circular Mongol home with a thirty-foot diameter described by William of Rubruck, its inch-thick felt covering and framework of wooden poles would have come to several tons; and the wheels, mast-like axle, and wagon structure would easily have boosted that to the neighborhood of eight tons.

Welcome as it is to find reliable pictorial corroboration of William of Rubruck's report on Mongol transport technology, albeit with horses pulling a building stone instead of oxen pulling a mobile home, the presence of such a large and utilitarian vehicle in eighteenth-century China makes one wonder why four-wheeled wagons did not become more popular. In China, as in Europe, the adoption of cannon eventually contributed to the military's use of four-wheeled vehicles. However, the elite classes in China were not as anchored in—or obsessed with— a tradition of knightly warfare as were the noblemen of Europe. The wheel technology was there, but the psychological disposition to adapt that technology to private elite transportation was not.

The history of wheeled transport in China began around 1200 B.C.E., when emperors of the Shang dynasty (ca. 1600–ca. 1046 B.C.E.) elected to ride in grand chariots fashioned after those used by the wagon nomads of Central Asia, with whom China was frequently at war (figure 86). As in the Mediterranean region, chariotry soon became a powerful force on Chinese battlefields. A line of chariot-borne archers with infantry advancing alongside made for a powerful army, but the army's command-and-control structure struggled to get the line to move forward in an even and disciplined fashion, and found it difficult to change its direction as the battle progressed. These limitations led to the improvement of infantry tactics and the mounting of archers on horseback, another innovation of China's steppe enemies. Thus chariot warfare became obsolete. Under the Han dynasty (206 B.C.E.–220 C.E.), generals still used chariots as command vehicles, but they did not ride them into combat.

Off the battlefield, men of rank continued on occasion to travel by chariot, but the practice died out well before the Mongol invasion of the thirteenth century. The reasons for its demise are not clear, but a shift in the definition of elite status from prowess on the battlefield to mastery of the Confucian classics probably played a role. A European nobleman imagined himself as a knight on horseback and expected to be tested in battle. Riding a horse spoke to his self-image. A young man aspiring to become a Chinese official was trained as a scholar and knew that he

FIGURE 86 A bronze model of a Chinese chariot, followed by an attendant, second century c.e. (DEA / E. LESSING / Granger, NYC—All rights reserved)

would be tested on his learning. How he traveled from place to place was a matter of secondary importance.

The best portrayal of Chinese transportation in the twelfth century comes to life in a panorama of a bustling city executed by the famous artist Zhang Zeduan (1085–1145). *Along the River During the Qingming Festival*, a seventeen-foot-long scroll painting, details the elegant culture of the Song period (960–1279), which immediately preceded the Mongol invasions (see figures 22 and 87). It depicts seventeen porters, ten palanquins borne by other porters, nine wheelbarrows, eight mule- or oxcarts, and a wheelwright's shop showing artisans assembling a large spoked wheel. Although some of the carts are quite elaborate, none are drawn by horses or resemble the aristocratic chariots of a thousand years earlier. A few men ride horses with attendants on foot, and one woman sits sideways on a horse being led by a man. But there are no four-wheeled vehicles.

Artists reimagined Zhang Zeduan's masterpiece more than once between the twelfth and eighteenth centuries. Of these, the Qing version of 1736, which includes the wagon carrying a massive building stone, seems to be the most detailed and elaborate (see figure 85). In addition to that vehicle, drawn by twenty-two horses or mules, the scroll depicts six other four-wheeled vehicles, none of which appears to have a pivoting front

FIGURE 87 Wheelbarrows, palanquins, porters, and pack animals crossing a bridge, as depicted in a detail of Zhang Zeduan's scroll painting *Along the River During the Qingming Festival* (twelfth century). (Collection of the Palace Museum, Beijing. HIP / Art Resource, NY)

FIGURE 88 Wheelbarrows, porters, riding animals, and a four-wheeled wagon, as depicted in a detail of the eighteenth-century version of Zhang Zeduan's scroll painting *Along the River During the Qingming Festival* (twelfth century). (Collection of the National Palace Museum, Taipei, Taiwan)

axle (figure 88). Four are pulled by two animals, either mules or oxen, and carry bulky loads. Unlike all the other vehicles in the painting, these four appear to have solid wheels, which gives them a distinctly rustic look. Another conveyance, with spoked wheels and a roof, carries both passengers and cargo. One might compare it with a stagecoach were it not for the fact that its three oxen, yoked two in front of one, must have traveled very slowly. The final vehicle is a toy children's wagon.

To get an overall sense of the relative frequency of various means of land transport in mid-eighteenth century China, one should compare these seven four-wheeled vehicles with the other modes of transport shown on the scroll: thirty-eight riders of horses or donkeys (excluding a group of recreational riders cavorting in the rural background), at least thirty-six porters, twenty-five palanquins borne by other porters, twenty wheelbarrows, nine carts drawn by oxen or horses, and three sedan chairs carried between horses. Given that the scroll produced in the eighteenth century is twice the length of that painted by Zhang Zeduan in the twelfth century, the tallies of transportation from the two are proportionally quite similar. Human muscle power in the form of porters carrying packs and bearing palanquins far exceeds animal power, and wheelbarrow pushers and pullers predominate in the depiction of wheeled vehicles.

The presence of four-wheeled vehicles in the eighteenth century and their absence in the twelfth century, along with the twenty-two horses or mules harnessed to a single wagon in the style described by William of Rubruck, strongly suggests that the four-wheel concept came to China during the Yuan dynasty, established by Genghis Khan's grandson Kublai Khan in the thirteenth century, more than a hundred years after Zhang Zeduan painted his scroll. But four-wheeled wagons were not adopted in great numbers and were used for only basic transport, not as private carriages.

As for an association of wheels with gunpowder, the Mongols' seizure of the fortress city of Xiangyang in 1273 sealed the fate of the Song dynasty. The two sides pummeled each other with "thunder-crash bombs," but these cast-iron vessels filled with gunpowder were delivered by catapult rather than by cannon. Only under the Ming dynasty (1368–1644), which expelled the Mongols from China, did cannon find a prominent place on the battlefield. Some were mounted on four wheels, and some on two. In the sixteenth century, the Ming generals came up with a variation on the Hussites' tactic of placing light cannon and musketeers in wagons, but they arranged them in a square rather than a circle.

What China lacked when it came to popularizing four-wheeled transport was Matthias Corvinus. Psychologically speaking, for the new four-wheel technology to have captured the imagination of the Chinese elite the way the Hungarian coach did that of the European aristocracy, someone of stellar rank and visibility—a Ming emperor, for example—would have had to set the style, as did the charismatic king of Hungary. This is why the portrayal, in an eighty-five-foot-long scroll painting called *Departure Herald*, of a Ming emperor traveling with a four-wheeled portable palace whose design is based on Mongol nomads' mobile homes challenges my theory that even after four-wheel transport was adopted from the Mongols, it remained a humdrum technology that was used mostly for hauling heavy loads.

Departure Herald depicts a vast imperial procession traversing a somewhat abstract countryside. The cavalcade becomes grander and grander as the viewer scans it from left to right, and it culminates with the person of the emperor, portrayed larger than any other figure, riding a black horse with plumes rising from the top of its bridle (see figure 92). Right behind the emperor comes a palace on wheels pulled Mongol-style by twenty-three horses harnessed ten abreast in two ten-horse rows, with an additional trio hitched directly to four wooden beams extending from the front of the palace (figure 89). As with the four-wheeled wagon transporting a huge building stone depicted in the eighteenth-century scroll painting, the harnessing is accomplished by means of ropes fastened to either side of the horses' collars.

Does *Departure Herald* not signify imperial endorsement of the idea of a four-wheeled mobile dwelling along the lines of those that so impressed William of Rubruck? And might it not, therefore, have conferred imperial favor on the use of grandiose four-wheeled vehicles by the Chinese elite? The answers to these questions depend on the answers to three others: Where is the procession going? What is the purpose of the mobile palace? And who is the emperor?

The procession involves hundreds of horsemen and personnel on foot, many of them in military costume, particularly those in closest

FIGURE 89 A four-wheeled portable palace, as shown in a detail of the scroll painting *Departure Herald* (ca. late sixteenth–early seventeenth century). (Collection of the National Palace Museum, Taipei, Taiwan)

proximity to the emperor. Amid this throng are a number of vehicles of increasing grandiosity. From left to right, the parade includes

- Two wheelbarrows festooned with pennants.
- Three two-wheeled carts, also decorated with pennants, carrying plain rectangular containers and pulled by seven horses, harnessed four in front of three.
- One plain tent-like pavilion, approximately fifteen feet long by seven feet wide, suspended like a palanquin between four horses.
- Three identically and magnificently ornamented shrine-like structures, roughly four feet square and ten feet high. The first is mounted on a two-wheeled platform pulled by four elephants, harnessed three in front of one (figure 90). The second, also on a two-wheeled platform, is pulled by seven horses, harnessed five in front of three. Huge banners extend from the back of both shrines. The third is mounted on poles

FIGURE 90 A shrine drawn by four elephants, as shown in *Departure Herald*. (Colection of the National Palace Museum, Taipei, Taiwan)

that are carried on the shoulders of twenty-eight bearers. The numerous attendants accompanying the shrines include two pairs carrying what look to be staircases to be attached to the shrines once they are put on the ground. But the shrines do not have any visible doors.

- Three large coffers, two of them draped, each carried by a dozen bearers.

- A cavalcade of more than a hundred riders in military raiment preceding six saddled horses with plumes on their bridles, each being led by an elegantly costumed groom.
- The mounted emperor, drawn larger than any other person in the procession and turned in his saddle to face the viewer. Two attendants follow on foot and hold parasols over his head.
- Twenty or so mounted soldiers.
- The mobile palace itself, approximately thirty feet long and twelve feet wide, resting on four wheels about four feet in diameter. There is no indication of a pivoting front axle. Banners angle upward from the back porch of the palace.
- Seven plain two-wheeled carts with small enclosed chambers, each pulled by four horses, harnessed three in front of one and driven by drivers in humble dress. These plain carts recall William of Rubruck's report that each large mobile home belonging to the wife of a Mongol leader was trailed by a hundred or more small covered carts.

A capsule description of the scene depicted in *Departure Herald* contains scant information:

> In this handscroll is a great imperial procession making its way to pay respects at the imperial tombs. Departing from the Te-sheng (Victory) Gate of the Peking [Beijing] city wall, the artists here depicted shops along the way and the appearance of ceremonial guards to the final destination of the imperial tombs, the final resting place for Ming dynasty emperors 45 kilometers [28 miles] from the capital at Mt. T'ien-shou.[2]

With regard to the itinerary of the procession, the Te-sheng Gate, shown at the right end of the scroll, still exists and is readily identifiable, as is the distinctive series of archways and gates leading to the Ming tombs tucked into mountain valleys on the far left of the scroll. The route includes a number of narrow, hump-backed bridges; several similarly narrow gates and arches; and a final sacred road, not shown on the scroll, lined with huge animal sculptures. These features raise the question of whether the palace as shown could actually have made the twenty-eight-mile journey. The horses harnessed ten abreast would have required a constant road width of thirty feet or more in order to pull the palace, and the lack of steering would have militated against any but the most gradual of curves.

Should *Departure Herald*, then, be regarded as imaginary? And if so, what is the function of the mobile palace? While it is highly decorated, it is actually not very large. The circular Mongol dwelling thirty feet in diameter described by William of Rubruck would have had twice as much interior space. Rather than being circular, the mobile palace consists of transverse rectangular sections with gabled roofs in front and rear, linked by a narrower middle section with a lower, barrel-vaulted roof. The front section has an entrance door off a walkway, and the rear section has a partially walled back porch.

This architecture bears some resemblance to an underground palace uncovered in the 1950s in the only one of the Ming tombs to be excavated in modern times (figure 91). A transverse rear chamber is approached through a tunnel and two longitudinal chambers. All three rooms are roofed with barrel vaults. There are also two side chambers, but they were never used. The underground palace lies beneath the tomb of the Wanli Emperor (r. 1572–1620), so the vehicle depicted on the scroll may conceivably be a model of it. But the Wanli Emperor died well after the reigns of the Xuande Emperor (r. 1425–1435) and the Jiajing Emperor (r. 1521–1567), the two rulers for whom, or in memory of whom, *Departure Herald* is supposed to have been painted. Could the mobile palace have been an earlier version of underground palaces that still lie undisturbed beneath the Ming tombs? Or, rather than being buried, could it have been built as a cenotaph—literally, an empty tomb (from the Greek *kenos* [empty] and *taphos* [tomb])—and *Departure Herald* painted to commemorate a procession to deliver the structure to the grave site of the emperor?

Alternatively, the scene may be an imagined vision of an emperor's journey into the afterlife, with the palace his intended residence there,

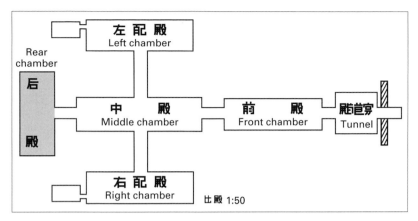

FIGURE 91 The plan of the palace found beneath the tomb of the Wanli Emperor.
(Author's collection)

rather than a trip to "pay respects at the imperial tombs." Three details seem to support this idea. First, the saddled and plumed horses being led in front of the emperor are probably his alternative mounts, perhaps to be ridden in the afterlife. Second, unlike all but a few of the hundreds of other horses depicted on the scroll, the horses pulling the palace are pure white, which is the color of mourning in Chinese culture. And third, the Eurasian steppe tradition of vehicle burial, which dates to the third millennium B.C.E., had evolved by the time of Genghis Khan into the secret interment of rulers along with the maintenance of a home aboveground at a nearby location. To this day, pilgrims pay pious visits to the cenotaph-temple dedicated to Genghis Khan in Inner Mongolia, while his actual burial site remains unidentified.

Finally, there is the question of the identity of the emperor. The painting is traditionally assigned to the reign of the Xuande Emperor. Like most of the Ming emperors, he was heavily bearded, but the emperor on horseback is clean shaven (figure 92). This has led some scholars to propose that the emperor depicted in the procession is the Jiajing

FIGURE 92 The emperor riding a black horse, as shown in *Departure Herald*. (Collection of the National Palace Museum, Taipei, Taiwan)

Emperor. However, most portraits of this monarch show him with a thin face rather than a round face and with a long but very wispy beard. In fact, the only Ming emperor who looks like the one portrayed in *Departure Herald*, and in the companion scroll *Return Clearing*, which shows an emperor traveling by boat back to the capital from the imperial tombs, is the Wanli Emperor, whose tomb is underlain by a palace. He has the requisite round face, and while some of his portraits show him with a wispy mustache, others depict him without facial hair. If this is the correct identification, then the palace on wheels may be viewed either as a spiritual representation of the palace beneath his tomb or as his intended home in the afterlife.

Returning to the question of why a grand painting of an emperor traveling with a four-wheeled mobile palace did not set a style for the Chinese elite, the answer is that the vehicle was part of a real or an imagined funeral procession. Far from suggesting that a high-ranking dignitary might show off his grandeur European-style by riding in a magnificent carriage, *Departure Herald* states that such grandeur may be necessary to an emperor's journey into the afterlife, but it is not for people still living in this world. Thus even though China adopted the technology of four-wheeled transport from the Mongols and, as in Europe, eventually transferred that technology into the military realm, no pattern of private use of four-wheeled conveyances by the civilian elite ever developed.

The question of psychology and elite preference arises in a different way in Japan. The Japanese have never taken transportation cues from China, but they may have from the West.

ten

Rickshaw Cities

The phenomenal success of the automobile gave birth to the notion that the wheel was humanity's greatest invention. It also sealed the fate of horses and other draft animals in the transportation arena. The animated film *Get a Horse!* (2013) imagines this historic moment. The bully Peg-Leg Pete, in his jalopy, yells at Mickey Mouse and his pals, who are riding in a hay wagon pulled by Horace Horsecollar, to "Make way for the future!" But the future was not limited to motor vehicles. Horsepower, in the literal sense, gave way to steam, gasoline, and electricity; but in so doing, it introduced a new era in human-powered transportation. Hundreds of millions of bicycles, shopping carts, baby strollers, dollies, gurneys, and roll-aboard luggage—none of which had been conceived of before 1800, but all of which are taken for granted today—moved into part of the space vacated by animal-drawn transport.

The first great success of modern human-powered locomotion was not the bicycle, which became a craze for athletically inclined young men with money to spend in the 1880s, but the rickshaw, a two-wheeled, hand-pulled passenger cart that was invented in Japan in or around 1869 (figure 93). Although the nineteenth century is famed for the invention in Europe and America of technologies that transformed

FIGURE 93 A pulled rickshaw in Singapore, 1901. (HIP / Art Resource, NY)

people's lives—not just trains and automobiles, but also steamboats, telegraph lines, and electric lighting—and many Europeans and Americans thought that bringing these blessings of the modern age to the backward peoples of the world justified Europe's imperial domination of the globe, the rickshaw was equally transformative in its impact on urban life and yet was born and developed exclusively in Asia.

The rickshaw (from the Japanese *jinrikisha* [human-powered vehicle]) provided a system of urban passenger circulation that was in certain ways superior to the options available in the West, but twentieth-century Europeans and Americans repeatedly expressed a visceral hatred of it. In their home countries, these detractors happily employed unskilled men and women at low wages to perform dirty and dangerous jobs in factories and mines, and, with a clear conscience, they tasked them with filthy chores like emptying chamber pots, sweeping chimneys, and collecting garbage. But a man standing between shafts to pull another adult in a light vehicle struck many of them as reducing the puller to the level of an animal. As American photographer Greg Vore, whose work *Rickshaw Wallah* documents the lives of rickshaw pullers in Bengal and Bangladesh, puts it: "What made the rickshaw so different from a wagon or an ox cart and in the eyes of many, so cruel, was the idea that it be pulled by a man instead of a farm animal."[1]

Pushing wheelchairs and baby carriages posed no problem for these critics, but *pulling* from the front—with the exception of the Bath chairs so popular at British seaside resorts (see figure 4)—turned a man into a beast. This quirk of perception reflects a history of human–animal relations that the Japanese—who ate little red meat, had few large herds of cattle and horses, and seldom used animals to pull vehicles—did not share with Westerners. Yet the Western viewpoint ultimately played an important role in suppressing the pulled rickshaw and replacing it with the cycle-rickshaw, also human powered, and the motor-driven auto-rickshaw.

Aside from one book in Japanese, *Jinrikisha* by Saitō Toshihiko, and two excellent studies of the impact of the rickshaw on specific places, *Rickshaw Coolie: A People's History of Singapore, 1880–1940* by James Francis Warren and *Rickshaw Beijing: City People and Politics in the 1920s* by David Strand, the history of the rickshaw has been minimally studied and is sometimes mired in claims and counter-claims as to who should get the credit for its invention.[2] No one disagrees on the date of invention, however, and no one denies that rickshaw use mushroomed almost overnight or that Japanese manufacturers produced tens of thousands of rickshaws for the local market and exported similar numbers to China and elsewhere in Asia.

As discussed earlier, Japanese society, unlike Chinese or Korean, made little use of wheeled vehicles before the mid-nineteenth century. The Tokogawa shogunate, Japan's central government, banned carts from major roads, reportedly to preserve the surface of the roads and protect the jobs of porters. Wheeled vehicles, such as the carts used to carry human waste from town latrines to suburban gardens and the shrines used in religious processions, thus served mostly local purposes. These vehicles as well as two types of cart occasionally used to transport freight, the *beka-guruma* and the *daihachi-guruma*, were pulled or pushed by humans.

Change began after foreign merchants were granted port and residential rights in Yokohama in 1859. At the urging of these merchants, who wanted to import carriages for their personal use, the shogunate began to relax its restrictions. In 1866, horse-drawn vehicles were permitted on the streets of Edo (present-day Tokyo), the capital, and on major roads. What followed was an explosion of vehicular experimentation (figure 94). Along with carriages and railroad cars carrying passengers in European dress, woodcuts from the 1870s display an astonish-

FIGURE 94 A Japanese woodcut showing an assortment of primarily human-powered vehicles (including rickshaws, a tricycle, and a paddleboat), ca. 1870–1900.

ing diversity of human-powered vehicles. Men in either European or Japanese clothing use pedals and cranks to propel land vehicles and paddleboats, but all the men pulling more practical and less experimental rickshaws and hand trucks wear Japanese attire. While these images may not precisely portray the vehicles of the day, they testify both to a lively spirit of experimentation at the beginning of the rickshaw era and to a focus on human muscle rather than animal power.

Nor did the arrival of the railroad in 1872 eliminate this focus. Several dozen narrow-gauge rail lines built between 1891 and 1910 took advantage of the reduced rolling resistance of iron wheels on iron rails, but for motive power used human muscle. The cars were pushed by hand (figure 95). The Japanese built many other hand-pushed railways for carrying freight and passengers in Taiwan, which they ruled as a colony from 1895 to 1945. One English visitor in 1920 described the ride:

> We were met . . . by the usual officials, who had push-cars and coolies ready waiting for us. The push-car is simply a light trolley with brakes. On the flat one coolie can get it along, running behind and shoving until it is well underway, and then getting on until it begins to slow down. Going up hill two coolies are necessary, but it is when you are going down hill that the fun begins and you have all the thrills of a prolonged journey on a kind of private (and rather flimsy) scenic railway.
>
> The track is a very light line of about 18 inches gauge and the sleepers are mere billets [short, thick pieces] of wood. At present only the main towns of [Taiwan] are connected by roads, and the outlying districts, even up to the hills, are linked up by means of these push-car lines, of which there are over 550 miles in the island. Whoever first thought of opening up the country in this manner was a genius, for in a land where there are few horses the push-car line is far more useful than a bridle-path would be, and moreover it forms an admirable method of transporting produce, such as camphor, from the hills.[3]

Automobiles aroused less interest than rail lines. The first Japanese motor vehicle, a steam-powered bus, appeared only in 1904, fully two

FIGURE 95 Men pushing a railway car with passengers on a Japanese narrow-gauge railroad. (Copyright © National Geographic Image Collection / Alamy)

decades after Karl Benz inaugurated the automobile industry in Europe. Thereafter, the Japanese military urged the manufacture of trucks, but not of cars. Not until 1917 did Mitsubishi produce its Model A, Japan's first mass-produced car. Based on a Fiat design, a total of twenty-two Model As were manufactured.

The influence on Japan of European modes of transport cannot be denied, of course, but experimentation with wheels was already under way before an American fleet commanded by Commodore Matthew C.

Perry, using blatant threats, forced the shogunate to open the country to the West in 1854. A woodcut by the artist Katsushika Hokusai, who died in 1849, shows an odd assortment of vehicles: men in a boat apparently using cranks to turn paddle wheels, a proto-rickshaw puller drawing a freight cart (probably a *beka-guruma*) with large spoked wheels, a passenger using sticks like ski poles to advance his little four-wheeled seat, a child's toy wagon, and a gang of men tugging on ropes attached to a huge wooden beam with a pair of solid wheels underneath it.[4] Again, the relationship of the image to real life cannot be determined, but the woodcut shows that humans moving things on wheels attracted at least some technical interest even before the arrival of European ideas.

The issue of inventiveness is important because of a historical dispute as to who invented the rickshaw. One camp sees it as the product of Yankee ingenuity and the propensity of the Japanese to copy other people's ideas. The other ascribes the invention to a trio of Japanese entrepreneurs. A third possibility is that several people came up with the idea more or less simultaneously in the context of the freewheeling experimental environment of the time. Indeed, if the report is anywhere near true that in 1872, only three years after the rickshaw's invention, the number in use had reached 40,000 and the vehicle had totally displaced the palanquin as the preferred mode of passenger transport in Japan, then the rickshaw design must have spread at almost miraculous speed, both to the workshops capable of building them and to the people who wanted to ride in them.

Supporting the notion that the rickshaw may have had multiple fathers is the fact that the three Japanese businessmen who applied for and received a rickshaw-manufacturing license in 1870 failed in their attempt to gain exclusive rights under Japan's first patent law, promulgated in 1871. Already there were too many workshops serving the burgeoning market. Two decades later, after a lobbying campaign to get pensions for the "inventors of the *jinrikisha*," the government's Awards Agency granted ¥200 (about $110) to Yasuke Izumi, the organizer of the enterprise, and the families of his two deceased partners.

In the meantime, in 1876, three months after an American consul general acquitted him for lack of a witness on the charge of having beaten a rickshaw puller who demanded immediate payment of his fee, Jonathan Goble, a one-time U.S. Marine turned Baptist missionary, petitioned the city government of Edo to grant him a share of the taxes generated by the city's 30,000 rickshaws. (The figure of 40,000 rickshaws in use in 1872 would thus seem to apply to the entire country.) The irascible Goble asserted that he had invented the rickshaw in 1869 so that his infirm wife could take the air and that he had taken a newly arrived, twenty-nine-year-old acting American consul to see his invention at the workshop of the Japanese artisan he had engaged to build it. In later years, the consul, who was reassigned to Yokohama after only a year, testified that the first rickshaw "was made under the direction of Jonathan Goble in 1869."[5]

One has to wonder, however, how an American diplomatic novice who lived in Japan's largest city for less than two years could have learned enough about what was being built in Edo's many workshops to be certain that Goble's rickshaw was the first. In the event, Goble's demand for a share of the tax revenue was rejected, and his story probably would have been forgotten if it had not subtly reinforced a stereotype of Americans as inventors and Japanese as imitators. To this day, historians speculate on whether Goble was copying an American baby carriage, some of which at that time had two wheels and shafts for pulling; the English Bath chair, which was pulled, but had three wheels and no shafts; or even the eighteenth-century French *brouette* (figure 96).

According to one of several conflicting stories, Goble had befriended a carriage maker from Pulteney, New York, when both of them were serving on Perry's flagship during the commodore's "opening" of Japan in 1854, and the wheeled chair that he designed was actually made in the United States and the parts sent to Japan for assembly. Unaccountably, few historians have noted the strong resemblance, particularly in the structure of the wooden wheels, between the rickshaws depicted in woodcuts from the 1870s and the farm carts and *beka-guruma* freight carts shown in woodcuts from before 1869.

Another mystery surrounds the equipping of rickshaws with elliptical leaf springs like those used in European carriages (figure 97). The first patent for a carriage mounted on springs of this sort was issued

FIGURE 96 Two servants pulling *brouettes* almost come to blows as the carriages block each other, in Claude Gillot's painting *Les deux carrosses*, ca. 1707. (Musée du Louvre, Paris. Gianni Dagli Orti / The Art Archive at Art Resource, NY)

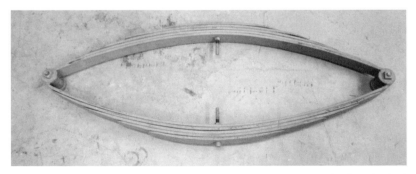

FIGURE 97 An elliptical leaf spring. (Photograph in author's collection)

in England in 1805, so it is not an independent Japanese invention. But when did it appear? By 1879, the man who would become Japan's most successful rickshaw manufacturer and exporter, Akiha (or Akiba) Daisuke, was certainly putting springs on his vehicles. But rickshaws had become the rage well before that date, and the early woodcut images do not show springs. Thus even though the importance of the springs cannot be overstated from a comfort standpoint, technological innovation may have contributed less to the initial success of the rickshaw than stylishness, just as aristocratic display rather than technology set the pace in the European carriage revolution.

Elite social status certainly played some role in the rickshaw's initial popularity, though the Meiji emperor himself, who presided over Japan's modernization during his forty-five-year reign (1867–1912), traveled everywhere in his domain in a royal palanquin and never rode on wheels. Akiha Daisuke's company catalog from 1911 contains an English-language description of those early years that clearly implies an association between rickshaw use and elite social status:

> The introduction of the Jinrikisha, in 1869, really open[s] a new era upon the mode of travel in Japan *where social status was still in its infant stage*. Soon after its debut, approbation followed appreciation, and its reputation spread like wild fire over the whole country, so that in the year 1872 saw [*sic*] the business growing in such flourishing a state as the customers often crowded in front of the store, and in some case, they were compelled to draw lot, in order to obtain a Jinrikisha. The same experience I had in Osaka, at the opening of my branch store in 1871. The customers were always impatiently waiting for the goods from Tokyo. Right on the arrival, all had gone leaving nothing at the store, even th[os]e damaged in the transit being gladly accepted.[6]

In "*Jinrikisha* in Meiji Japan," Lauryn Noahr supports this implication:

> The social elites favored the *jinrikisha* and it soon became a status symbol. . . . The wealthier Japanese families, officials, and businesses often

hired private *jinrikisha* men. The private *jinrikisha* were often decorated differently and had constant changes or perks to them. Some included the hood, to cover the rider when it was raining, sunny, or snowing; decorated sides, rubber wheels, and fancier uniforms for the *jinrikisha* men. The private *jinrikisha* owners started many trends which trickled down to the lower classes and quickly became must-haves for them. The rubber wheels are one such example.[7]

The most popular conveyance for men and women of status before the rickshaw was invented, a type of palanquin called a *kago*, quickly disappeared from the streets of Edo, even though it weighed very little, was easy to get into, and bounced minimally because the legs of the two bearers effectively functioned as shock absorbers (figure 98). Riders might

FIGURE 98 A woman being borne in a traditional Japanese palanquin (*kago*), 1880s.

sway; but no matter how irregular the road surface, they suffered few hard jolts. Trading this relatively smooth ride for a vehicle that perched its passenger directly over the axle must have marked a major decline in comfort. But style often wins out over comfort, and the quick adoption of elliptical leaf springs of the type used on the carriages imported by Europeans made for a markedly superior ride.

Social status tied to a "rickshaw revolution" seems to have played a lesser role in the spread of rickshaws equipped with elliptical leaf springs and folding tops outside Japan. The ways in which the new vehicles transformed urban environments in East, South, and Southeast Asia, especially in the port cities of the Indian Ocean, can most easily be addressed under three headings: economic efficiency, urban lifestyle, and labor force.

Economic Efficiency

First, the rickshaw required just one puller rather than the two bearers needed to carry a *kago*. Given that the skill levels were more or less equivalent, a rickshaw rider could travel twice as far for a given fare.

Second, the rickshaw was cheap to build and maintain. This made it more efficient than comparable one-passenger vehicles for hire in European and American cities. The hansom cab of London, to take one possible comparison, was much more expensive to build and required both a horse and a driver (figure 99).

Third, rickshaw operators, who often owned substantial fleets of vehicles, incurred no horse-related costs: maintaining harnesses and feeding, stabling, shoeing, and eventually replacing animals.

Fourth, the rudimentary skills required of a rickshaw puller and the abundance of unskilled immigrants from the countryside and abroad who regarded the job as an entryway to the life of the city allowed the wages of a rickshaw puller to be much lower than those of a carriage driver, who had the ability to manage one or more horses on city streets.

FIGURE 99 A hansom cab in London, 1855.
(Photograph © Andrew Dunn, September 19, 2004)

Urban Lifestyle

First, human muscle power meant no horse manure on the streets and no in-town stabling and feeding of horses. In Japan, rickshaw pullers even wore uniforms that suggested they were part of a modern way of life.

Second, rickshaws took up little space on the street. The footprint of a rickshaw (approximately 3 × 5.5 feet) was smaller than that of either a hansom cab (approximately 4 × 20 feet) or an automobile (Model T Ford [5.5 × 11 feet], Honda Civic [5.5 × 15 feet], Smart [5 × 9 feet]). When not in service, rickshaws could be nested together, thereby reducing the need for parking space (figure 100). Having two wheels also diminished traffic congestion, since rickshaws could make sharper turns

FIGURE 100 Nested storage of rickshaws. (© Brett Cole)

than any four-wheeled vehicle, and pullers could maneuver more finely than any carriage driver.

Third, on flat ground the pullers commonly ran rather than walked, so passengers could travel twice as far in a given amount of time as they could on foot or in a palanquin. This effectively increased the radius of central cities that had evolved historically on a pedestrian scale and contributed to the characteristic bustle of modern life.

Fourth, during the monsoon season in cities like Kolkata (Calcutta), rickshaws were able to function in flooded streets that motor vehicles could not negotiate.

Fifth, rickshaw men, especially those who pulled privately owned vehicles, took over many functions traditionally performed by household servants, such as taking children to and from school and running errands.

Labor Force

The rickshaw arrived in Shanghai in 1874. By the end of that year, ten companies, all owned by Europeans but importing equipment from Japan, were offering rickshaw service. The number of vehicles peaked in 1930 at more than 23,000, or 1 public rickshaw for every 150 residents.

In Singapore, where Chinese merchants from Shanghai began to import rickshaws in 1880, the number in use reached almost 30,000 in 1923, or one for every fifteen residents. Most of the pullers were Chinese immigrants from Fujian Province.

Rickshaw use in Beijing, an inland city, began somewhat later, in 1886. In the 1930s, the number of vehicles soared to more than 40,000. As David Strand writes: "Rickshaw pulling was a public spectacle in Beijing in the 1920s. Sixty thousand men took as many as a half million fares a day in a city of slightly more than one million people."[8]

The thousands of unskilled workers who migrated to these cities to pull rickshaws, sometimes from nearby villages and sometimes from abroad, constituted a new proletariat. Being dependent on employers who owned fleets of rickshaws, they were ripe for labor organization, strikes, and mobilized opposition to perceived threats like the building of streetcar lines.

Thus the role played by rickshaw pullers closely resembles that played by unskilled mill workers in European and American cities, complete with abusive treatment by rickshaw owners and unsavory living conditions in residential warrens. Upwardly mobile rickshaw entrepreneurs normally owned a small number of vehicles, which they rented to pullers for a share of their earnings. Unlike in mills, however, child labor was not a problem because it took adult strength to pull a rickshaw, and working outdoors, at least in mild climates, was probably healthier than laboring in a factory.

The restructuring of the urban labor market lent a degree of similarity to cities like Tokyo, Beijing, Singapore, Kolkata and Mumbai (Bombay) in India, Colombo in Sri Lanka (previously Ceylon), and Durban in South Africa. Cities under French and Dutch colonial rule seem to have been less involved with Japanese export businesses than those under British rule. Hanoi (Vietnam) and Phnom Penh (Cambodia), both in French Indochina, and Jakarta in Indonesia had far fewer pulled rickshaws.

Although humanitarian concerns about the living conditions and "animalistic" abasement of rickshaw pullers converged with the development of other forms of public transit in achieving the abolition of the pulled rickshaw in most cities, the economy and efficiency of the rickshaw system has survived. At last report, some six thousand rickshaws are still plying the streets and lanes of Kolkata.[9] The future of the rickshaw may not lie with pulled rickshaws, however. The cycle-rickshaw, which can carry passengers either in front of or behind the operator (figures 101 and 102), exhibits many of the virtues of its predecessor, as, to a lesser degree, does the auto-rickshaw, which is equipped with a small (but environmentally polluting) internal-combustion engine. And the future of the rickshaw may not lie in Asia and the Indian Ocean region, where automobile ownership is steadily growing. Rather, the slow but steady growth of cycle-rickshaws of more or less modern design in the cities of Europe and the Americas attests to a belated appreciation in the West of this form of urban wheeled transport (figure 103).

The history of the rickshaw challenges our notions about how inventions come about. Although the rickshaw involves one technological innovation rooted in Europe's industrial revolution—the addition of elliptical leaf springs, which require high-grade steel—its enthusiastic reception in Japan preceded the addition of this feature to the basic design. Inasmuch as the rickshaw required no manufacturing skills beyond those readily available in a wheelwright's shop, the original design of a light, two-wheeled, hand-pulled passenger seat or cabin could have been hit upon at other times and in other places—indeed, was hit upon in the eighteenth-century French *brouette*. But no earlier version ever awakened

FIGURE 101 A cycle-rickshaw (*pouse-pouse*) in Vietnam, with the passenger seated in front of the operator. (Istock / tbradford)

FIGURE 102 A cycle-rickshaw in India, with the passengers seated behind the operator. (Istock / pius99)

FIGURE 103 A modernized cycle-rickshaw in Berlin. (Istock / Kerstin Wauri)

extensive consumer interest. So were rickshaws simply a local Japanese fad? No, they were not. Businessmen of different national backgrounds exported them from Japan, and they quickly became popular in many countries with highly diverse transportation traditions. Yet they were not initially imitated in the West, even though European businessmen imported them into Shanghai. Why, then, did they succeed so broadly, but not universally?

The book *Does Technology Drive History?* edited by Merritt Roe Smith and Leo Marx, brings together thirteen essays that explore the social and economic contexts of invention in different times and places.[10] Some contributors tie modern transformative inventions to the rise of a capitalist system in which market competition rewards innovation and entrepreneurial investment. They cite Karl Marx's declaration that "in acquiring new productive forces men change their mode of production; and in changing their mode of production, in changing the way of earning their living, they change all their social relations. The hand-mill gives you society with the feudal lord; the steam-mill society with the industrial capitalist."[11] From this perspective, precapitalist technological change sometimes appears to be more or less random, with no underlying driving force.

Other contributors emphasize social, political, and cultural contexts rather than the capitalist marketplace. Rosalind Williams, for example, writes:

> The appeal to technology as a revolutionary force is therefore not particular to Marxism. It is part of a comprehensive view of inevitable historical progress that emerged in the Enlightenment and still endures, though greatly weakened. Technological determinism is an integral part of that theory of progress, according to which technologies of communication and transportation will conquer not just the clergy and the aristocracy but history itself.[12]

None of the theories espoused in the book shed much light on the history of the rickshaw, however. On the one hand, the rickshaw is modern. It was invented and found immediate popularity in a nation that was experiencing rapid modernization in the mid-nineteenth century; it triggered innovations in labor recruitment and organization; it transformed the urban way of life in every country that adopted it; and it was marketed by Japanese, Chinese, Indian, and European businessmen who were well aware of modern manufacturing and advertising practices. Akiha Daisuke himself diversified, but did not abandon, his rickshaw production in 1914 by creating the Number One Auto Company and the Aoi Automobile Company as a representative of Ford and Chevrolet.

On the other hand, the rickshaw has no direct connection with the modern industries of Europe and America. Despite the claims of Jonathan Goble, its invention was not the result of an ideology of progress. It competed, and still competes, with the European transportation technologies of railroads, streetcars, and automobiles. And the armies of rickshaw pullers to which it offered low-paid urban jobs formed a proletariat that was decidedly nonindustrial.

The use of the rickshaw spread in societies where human muscle power was well established in the transportation economy. Neither wheeled vehicles nor the animals that pull them were widely employed in Japan, wheelbarrows were the most widely used wheeled conveyances in China, and all the countries that adopted the rickshaw already utilized palanquins and human porters. Europeans regarded all these forms of transportation as primitive, and they particularly derided what they thought of as the "bestialization" of rickshaw pullers. But rickshaws can also be regarded as local low-tech responses to the horse-drawn carriages that contributed to the lifestyle of European imperialism.

Confusion as to how to account for a new and transformative technology that was entirely non-European and seemingly quite old-fashioned may explain why historians have so rarely commented on the rise of the rickshaw. Yet the rickshaw is as much a part of the transportation revolution of the past two centuries as the railroad and the automobile, and it may still have an important future in Western cities in the form of cycle-rickshaws. Nor is it the only type of human-powered vehicle in today's horseless world—far from it.

The Third Wheel

The last quarter of the nineteenth century witnessed the spread of three revolutionary innovations in human-powered wheeled transport: the rickshaw, the bicycle, and the caster. The caster, the most important of these inventions, has never before been subjected to serious historical inquiry. None of the three can be regarded simply as a response to the disappearance of horse-drawn vehicles, because the roots of all three go back to before the steam locomotive first cast a shadow over the future of the draft horse. Nor were they offshoots of a single technological trend. Yet the near simultaneity of their rise to popularity seems too striking to be merely coincidental.

The caster is particularly mysterious because unlike the rickshaw and the bicycle, which were invented for transporting people, it apparently originated as a way of shifting furniture from place to place and only slowly was adopted for use on vehicles, starting, perhaps, with wheelchairs. Hence it is hard to link these three technologies to changes in social status, even though rickshaws and bicycles do seem to have such a connection.

Wheeled vehicles propelled by human muscle power had always played a minor role in Europe and America, and they continued to be used in specialized circumstances throughout the carriage revolution and the dawning of the age of motorized transport. Hand-pushed ore-cars remained indispensable to miners, wheelbarrows were a mainstay of home builders and gardeners, and nineteenth-century nannies and nurses took to pushing baby carriages and wheelchairs. Pushcarts and handcarts provided places of business for peddlers and street vendors, and they sometimes acquired a more heroic image, as in the many depictions of the Latter-Day Saints trekking to Utah (figure 104). On the whole, however, hand-pulled vehicles signaled hard work, economic hardship, or servile occupations.

FIGURE 104 C. C. A. Christiansen's romanticized portrayal of Latter-Day Saints using handcarts on their westward trek to Utah, in *Handcart Pioneers* (1900). (Collection of Church History Museum, Salt Lake City, Utah)

The bicycle presented a different image, for the faddish bicyclists of the 1880s came from the more prosperous ranks of society, like today's hipster riding his fixed-gear bicycle to Starbucks. Moreover, their equipment needs inspired inventors and tinkerers, some of whom went on to experiment with automobiles and airplanes. And being from the social elite, the bicyclists' lobbying prompted local governments to appropriate money for smoother street paving.

With its two wheels in line instead of side-by-side, as on a cart or rickshaw, the bicycle also utilized a new manner of steering. Riders could change direction by leaning, shifting their center of gravity to one side or the other. Some of the "ordinary," or "penny-farthing" (from the contrasting size of the huge British penny and the tiny British farthing), bicycles that predominated through the 1890s relied entirely on this sort of "indirect" steering and did not have pivoting front wheels (figure 105). The handlebars, wedged against the rider's thighs, were designed to secure his or her position high off the ground rather than to change the direction of the front wheel. In effect, the "ordinary" was a unicycle with a small trailing wheel to support the rider's rearward displacement of the bicycle's center of gravity. Expert riders leaned far forward to balance as much of their weight as possible directly over the front axle (figure 106). Steering by leaning had long been used by ice-skaters and skiers, of course, but now it entered the post-equine world of wheels and even became an essential aspect not just of bicycling but also of motorcycle riding, in-line skating, and skate-boarding.

Bicycles appeared on city streets in Japan, India, and China around 1890, a decade and more after those countries adopted the rickshaw. Whereas rickshaw pullers came from an exploited underclass, expatriate Europeans and local youth who felt the attraction of European styles pioneered the new device, as much for sport as for transportation. In keeping with the identification of velocity with masculinity in the late nineteenth century, men in Europe and America began competing in bicycle races as early as 1870, and cycling became a staple of the Summer Olympic Games from their inauguration in 1896. Not until the 1940s, however, did various forms of bicycle racing, both closed-track

FIGURE 105 A "penny-farthing," or "ordinary," bicycle in the United States. (Copyright © Archive Pics / Alamy)

and cross-country, excite similar interest in Asia, where they became a venue for betting. Thus aside from being powered by human muscles, the bicycle and the rickshaw followed quite different paths, at least until cycle-rickshaws began to show up in one country after another beginning in the 1930s.

Just as the trajectories of the bicycle and the rickshaw do not overlap, neither do those of the bicycle and the caster, except at the most basic conceptual level. The bicycle evolved from the velocipede, which

FIGURE 106 Turning by leaning in a penny-farthing bicycle race.
(© Stephen Chung / Alamy)

was also known as the hobbyhorse or dandy horse because of the foppish manners of its devoted young riders. Invented by Baron Karl Drais in Mannheim, Germany, in 1817, the velocipede—his version was called a *Laufmaschine* (running machine) or, after him, a *Draisine*—had two wheels of similar size and looked like a bicycle without pedals. The rider straddled the seat and used his feet to propel the vehicle forward as if he were walking or running. (A French inventor came up with the idea of attaching pedals to the hub of the front wheel in 1863.) As with all modern bicycles and some "ordinaries," the front wheel of the *Draisine*, though not of all velocipedes, turned left or right around an axis of pivoting that was offset by six inches or so from the wheel's axis of rotation. Technically, this made the wheel of the *Draisine* a caster.

The caster emerged in the eighteenth century as the first new wheel concept since the fourth millennium B.C.E. Alongside the wheelset, whose wheels are fixed to the ends of an axle and turn together as a wheel-axle-wheel unit, and the independently rotating wheels at either end of an axle, the caster embodies a third wheel idea: a single wheel that can both roll and pivot (figure 107). It offered a new solution to the age-old problem of how to steer, and it thereby revolutionized human-powered vehicular transport. Although it was invented at least a century before the velocipede, in the form of tiny wheels attached to the legs of pianos, chairs, and other pieces of furniture to make them easier to move, the principle that the caster exploited was the same as that found in Drais's velocipede. When the horizontal axis of rotation of a caster, or

FIGURE 107 Antique furniture casters.

THE THIRD WHEEL

FIGURE 108 A motorcycle with an extreme caster angle. (Flickr/Michael Pereckas)

of the front wheel of a bicycle or motorcycle, is offset before or behind the vertical axis on which it pivots, any force coming from the side causes it to swivel automatically into alignment with the direction from which the pressure is being felt. The degree of the offset between the two axes is called the caster angle or the trail (figure 108). The trail is what makes "no-hands" bicycle riding possible because the lateral pressure exerted by the rider leaning to the side causes the front wheel to self-correct its steering. However, a minimal trail, or no trail at all, gives the rider greater direct steering control.

The caster angle of a bicycle or motorcycle affects the rider's stability and control, but casters function differently on three- or four-wheeled vehicles. A heavy piece of furniture, a laundry cart, a dolly, and a desk chair—to take some of the purest forms of caster use—can be pushed or pulled in any direction without resort to direct manipulation of their wheels. Vehicles like shopping carts, baby strollers, and wheelchairs that have one or two casters in front of or behind two nonpivoting wheels can be steered by twisting the handle(s) from behind. In both arrangements, the caster provides an ease of steering unattainable in any animal-drawn vehicle. Steering becomes a nightmare, however, when a caster fails to pivot, as with a malfunctioning shopping cart.

Did Drais incorporate a caster angle in his design for a velocipede because he understood how casters worked on furniture or, perhaps, on the first caster-equipped wheelchairs, which date from the eighteenth century? Or was his an independent invention? Since the origin of the caster is virtually unexplored, there is no current answer to this question. Indeed, the most common story of the caster's invention is untrue. David A. Fisher Jr., an African American working in a furniture factory in the District of Columbia, is commonly credited with inventing the caster in 1876. He is also considered the inventor, in 1875, of the joiner's clamp, an invaluable device used by woodworkers to hold glued joints in place while the glue dries. On the basis of these achievements, Fisher might well be considered the most important black American inventor of the nineteenth century. A reading of his patents, however, reveals more modest, though still valuable, breakthroughs. One patent is headed "Improvement in Joiners' Clamps" and the other "Improvement in Furniture-Casters," and both contain language to the effect that joiners' clamps and furniture casters were commonly in use at the time that Fisher's patents were issued. The caster patent, in particular, states: "The object of my invention is to provide a caster for furniture, &c., in which the caster wheel with its shank or spindle [the pivot] will not fall out from its socket as easily as *those now in ordinary use*, and which can be readily applied."[1]

This takes Fisher out of the picture as the inventor of the caster, but it leaves open the question of who did invent it. If the caster was already in ordinary use in 1876, when and where did it originate? This question may seem trivial or inconsequential until one considers that every dolly, gurney, shopping cart, laundry cart, baby stroller, wheelchair, and rolling desk chair—not to mention some assortment of suitcases being maneuvered through airports—operates on casters. Indeed, casters are part of the life of virtually every American from the cradle to the grave. A newborn is rolled to the nursery in a caster-mounted bassinet, and a corpse is rolled to the mortuary on a caster-mounted gurney.

In terms of chronology, the word "caster" (sometimes spelled "castor") already appears in 1828 in Noah Webster's dictionary of American English with the following definition: "A small wheel on a swivel, on which furniture is cast, or rolled, on the floor." The word is not included, however, in Samuel Johnson's famous dictionary of 1755. The authoritative *Oxford English Dictionary*, which began publication in 1884, cites it under the spelling "castor" and provides an example of usage from 1798. In material terms, antique dealers date English furniture casters with wooden wheels to the period 1720 to 1760, if not earlier, with cast-iron and brass casters following in the second half of the eighteenth century. All this evidence, of course, pertains to tiny wheels placed under the legs of furniture. For the conceptual transfer to larger wheels suitable for other purposes, the earliest surviving image, from 1766, is in a Finnish source and shows a caster-equipped wheelchair (figure 109).

This conceptual leap was far from obvious and took a long time to gain headway. When James Heath invented the Bath chair some twenty years later, he did not offset the rotational and swiveling axes of its small front wheel. This made the steering of the chair dependent either on a puller or on a passenger manipulating a tiller (see figure 4). Heath's company was still advertising chairs of this design in the mid-nineteenth century, and they continued in use with their nattily attired attendants into the early twentieth century.

On the heavy-duty side, what were called factory carts or factory trucks in 1900 seem to be the ancestors of today's dollies. How the word "dolly" arose remains unknown, but general factory applications surely evolved from furniture manufacturers, which, as we have seen in regard to Fisher's patent, were utilizing casters in factory settings by 1876. In the early twentieth century, factory carts typically combined a pair of large diameter fixed wheels in the middle with either one or two casters before and behind them to facilitate steering (figure 110). The all-

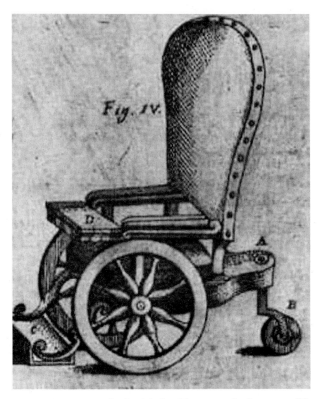

FIGURE 109 A wheelchair with a caster in the rear, 1766.
(Photograph in author's collection)

caster dolly presumably developed afterward, but evidence for this is
lacking.

This history of 5500 years of wheeled transportation began with a question: Why invent the wheel? It will close with another question: Why
invent the caster? Just as people managed for tens of thousands of years

FIGURE 110 A factory cart with a wheelset in the middle and casters at each end, early twentieth century. (Photograph in author's collection)

to transport themselves and their goods without wheeled vehicles, and often made little use of wheels even after they became aware of carts and wagons in neighboring lands, so people got along quite happily without casters for more than 5000 years after the wheel was invented.

Making small wooden wheels would never have posed a technical challenge, and by the first millennium B.C.E., neither would crafting them out of bronze or iron. The same can be said for the curved metal bracket that fastens at one end to the caster's short axle and terminates at the other end in a shank or spindle designed to pivot in a socket. Nor would providing a socket in which the spindle could turn have proved difficult, since metalworkers had been designing wooden hilts to receive the tangs of swords and metal tools for centuries, and animal fat made an effective lubricant. So the caster could have been invented in ancient or medieval times in any one of the Old World's wheel-using societies:

Mesopotamia, Egypt, China, Central Asia, Europe, India, Southeast Asia, or elsewhere. But it was not.

The caster's apparent invention as a means of moving furniture hints at two considerations that may have retarded its invention. First, although there is no intrinsic limit to the size of a caster—as we have seen, the front wheel of most bicycles can technically be considered a caster—for moving a heavy load, as opposed to helping a bicycle rider maintain stability, the socket in which the spindle or tang of a caster pivots should be situated on top of the caster and underneath the load. This poses a problem. For centuries, most carts, wagons, and carriages had wheels that were thirty inches or more in diameter. The bed of the vehicle, whether bearing passengers or freight, sat much lower, however, just above the axle of a cart or above the rear axle of a wagon or carriage—that is, at a level just over half the diameter of the vehicle's wheels. If a wagon had thirty-inch casters instead of normal wheels, the passengers or freight would be situated more than twice as far off the ground, for the wagon's bed would have to incorporate or go on top of the sockets holding the pivoting spindles, just as in dollies, gurneys, and laundry carts. A passenger or freight wagon, as opposed to these smaller and lighter vehicles, would thus have had such a high center of gravity that it would have been in danger of tipping over, particularly since it could so easily slip sideways.

Second, furniture casters are seldom more than two inches in diameter, and the objects that roll on them are large, oddly shaped, and sometimes very heavy. Therefore, they require a smooth surface to move on. Tiny wheels cannot go over bumps, cracks, holes, or obstacles. The mosaic floors of the Roman era or the majolica-tile floors of the Renaissance may have been smooth enough for casters, but they were rare and expensive. As for operating outdoors, furniture casters could not have navigated the bricks, cobblestones, and wooden setts that made for the smoothest paving before the mid-nineteenth century. Hardwood floors in a living room or a furniture factory would have provided the only commonly available smooth surfaces. Hence the caster's seeming origin as a way to move furniture and its association with indoor settings.

Today's casters seldom have a diameter of more than five inches, though wheelchair casters may go up to eight inches or so, and baby strollers for jogging parents may have a twelve-inch wheel. Thus the transfer of the caster concept from moving furniture to more diverse applications, such as wheelchairs, can reasonably be related to the availability of smoother, cheaper, more durable, and easily replaceable indoor flooring materials like linoleum, which was invented in 1855, and vinyl tile, which came along in 1933. Outdoors, asphalt and concrete provided similar improvements, though both are subject to cracking and crumbling. Nevertheless, smooth floors made of wood or ceramic tile long predated the modern era, so it is puzzling that no one thought of mounting an indoor object on casters over the thousands of years that elapsed after the invention of the first two types of wheel.

Unrestricted lateral movement, despite its convenience in turning and when pushing a load sideways, raises three additional complications. First, taking a dolly as an example, if the load on the dolly is heavy and the surface it is rolling on slants to the side, gravity may cause the dolly to slip sideways. Second, if a dolly is being pulled, the puller facing forward may not be able to correct for an unexpected sideways movement. And third, a lateral push that is either too far forward or too far back can cause the dolly to skew at an undesired angle. To obviate these difficulties, the casters on factory carts around 1900 were usually placed both in front of and behind a wheelset of larger diameter, and shopping carts and wheelchairs operate with casters in front and independently rotating wheels in back. In addition, today's casters usually have a lock to prevent unwanted movement, whereas casters of old were placed in cups once a piece of furniture was positioned, both to prevent movement and to preserve a floor or carpet from indentation.

A hypothesis that may be drawn from this is that the invention of the caster derived from trying to improve vehicular steering by combining a pivoting wheel with either wheelsets or independently rotating wheels, as in shopping carts and wheelchairs. But this hypothesis cannot account for the apparent fact that the caster first became common as a device for moving furniture. If we take a step backward, however, and look at the spread of the caster in a broader historical context, it will be useful to recall the many times that historians and archaeologists, à la Jared Diamond, have expressed astonishment that the indigenous peoples of the Western Hemisphere never thought of the wheel as anything more than a child's plaything and have attributed this lack of vision to the absence of domestic draft animals in their societies. I have argued in this book not only that the first wheeled vehicles in Europe were pushed by miners rather than drawn by oxen, but also that vehicles propelled by humans can be both economically efficient, witness China's wheelbarrows and Japan's rickshaws, and ritually satisfying, as with Japan's *dashi* floats and India's juggernauts.

Today, again, domestic draft animals are largely absent from our transport systems. We have them, but we no longer make use of them. The steady disappearance of horses and oxen from roads, streets, and farms over the past century and a half has coincided with the renaissance of small-scale human-powered transport. Casters have never been used in conjunction with animal power because harnessed animals do not exert lateral force. Nor can they exert torque by pushing harder on one handle, or on one side of a single handle, than on the other. Draft animals either push or pull. Any other application of force is irrelevant.

To reverse the spurious argument concerning the absence of domestic animals in pre-Columbian America, we might consider whether the horse and the ox prevented people in the Old World from experimenting with human power. As long as a mind-set fixed on animal traction dominated wheeled transportation, the evidence for which can be found in the association that Europeans and Americans so often make between rickshaw pulling and bestial servitude, thinking outside that mind-set was as difficult and unlikely as a pre-Columbian potter looking at a toy dog on wheels and imagining a wagon or a cart.

Casters first came to light in places where draft animals never trod, drawing rooms and furniture factories, and where loads were simply manhandled—forward, backward, sideways—without any pulling or pushing in harness. As soon as the idea verged on pulling and pushing, as it did with the Bath chair, the draft-animal mind-set took over. The maneuverability of the caster marked a major advance in controlling the direction of heavy things rolling on three or four wheels, but it could not be used on vehicles with two wheels placed side by side, such as the hand truck, or on the one-wheeled wheelbarrow.

Horses still predominated on streets and roads at the end of the nineteenth century, but plausible visions of self-propelled automobiles spread rapidly as the wave of the future quickly broke through the draft-animal mind-set in the first decades of the twentieth century. Carriage makers remained in business for a while, but the clever young mechanics who invented motorcars and airplanes, men like Henry Ford and Orville and Wilbur Wright, got their start exploring human muscle power in bicycle workshops. Even in Japan, where animal traction had never become firmly entrenched, the advent of European transportation notions helped spur a wave of experimentation that culminated in the rickshaw.

It seems unlikely that the caster was invented just once in some unidentified English furniture factory, just as it seems unlikely that wheelsets and independently rotating wheels were invented only once. Invention leaves little or no trace, however, if it is not accompanied by adoption. And adoption, as we have seen, is affected by many considerations. Some are broad, like economic efficiency, military utility, social class, gender, aesthetics, and religion; some are local, like availability of wood and roughness of terrain. In illuminating the manifold interconnections among these various considerations, the story of the wheel helps us understand that invention is seldom a simple matter of who thought of something first.

Notes

1. Wheel Versus Wheel

1. William Gilchrist, "The Carriage Tax Considered in Reference to Taxes on Locomotion, and the Influence of Taxation on British Carriage Building," in *Papers Read Before the Institute of British Carriage Manufacturers, 1883–1901* (Aspley Guise: Powage Press, 1902), 21.
2. [John Kitto], "The Deaf Traveller—No. 5. Vehicles of Persia and Turkey," *Penny Magazine of the Society for the Diffusion of Useful Knowledge*, October 19, 1833, 407. I am grateful to Dr. Ariel Salzmann for this reference.
3. Samuel G. W. Benjamin to Frederick Theodore Frelinghuysen, Tehran, October 2, 1883, Despatches from United States Ministers to Persia, 1883–1906, Diplomatic Series no. 28, United States National Archives, quoted in Shiva Balaghi, "Nationalism and Cultural Production in Iran: 1848–1906" (Ph.D. diss., University of Michigan, 2008), 1. I am grateful to Dr. Hossein Kamaly for this reference.
4. M. G. Lay, *Ways of the World: A History of the World's Roads and of the Vehicles That Used Them* (New Brunswick, N.J.: Rutgers University Press, 1992), 67.
5. Ibid., 68.
6. James Sterling Young, *The Washington Community, 1800–1828* (New York: Columbia University Press, 1966), 75.

7. Ibid., 76.

8. Steve Kemper, *Reinventing the Wheel: A Story of Genius, Innovation, and Grand Ambition* (New York: HarperBusiness, 2005).

9. Wolfgang Schivelbusch, *The Railway Journey: The Industrialization of Time and Space in the Nineteenth Century* (Berkeley: University of California Press, 1986).

10. Quoted in Thomas Zeller, *Driving Germany: The Landscape of the German Autobahn, 1930–1970* (New York: Berghahn, 2006), 138.

2. Why Invent the Wheel?

1. Jared Diamond, *Guns, Germs, and Steel: The Fates of Human Societies* (New York: Norton, 1997), 248.

2. David W. Anthony, *The Horse, the Wheel, and Language: How Bronze-Age Riders from the Eurasian Steppes Shaped the Modern World* (Princeton, N.J.: Princeton University Press, 2007), 72 (emphasis added).

3. Richard W. Bulliet, *The Camel and the Wheel* (Cambridge, Mass.: Harvard University Press, 1975).

4. [John Kitto], "The Deaf Traveller—No. 5. Vehicles of Persia and Turkey," *Penny Magazine of the Society for the Diffusion of Useful Knowledge*, October 19, 1833, 407.

5. Alan Macfarlane, "The Use of the Wheel in Japan," 2002, www.alanmacfarlane.com/savage/A-WHEEL.PDF (accessed October 8, 2014).

6. J. W. Robertson Scott, *The Foundations of Japan: Notes Made During Journeys of 6,000 Miles in the Rural Districts as a Basis for a Sounder Knowledge of the Japanese People* (London: Murray, 1922), 49.

3. A Square Peg in a Round Wheel

1. Stuart Piggott, *The Earliest Wheeled Transport: From the Atlantic Coast to the Caspian Sea* (Ithaca, N.Y.: Cornell University Press, 1983), 63.

2. Ibid., 33.

3. Mária Bondár, "A New Copper Age Wagon Model from the Carpathian Basin," in *Archaeological, Cultural and Linguistic Heritage: Festschrift for Erzsébet Jerem in Honour of Her 70th Birthday*, ed. Peter Anreiter, Eszter Bánffy, László Bartosiewicz, Wolfgang Meid, and Carola Metzner-Nebelsick (Budapest: Archaeolingua Alapítvány, 2012), 78 (emphasis added).

4. Peter Stadler et al., "Absolute Chronology for Early Civilizations in Austria and Central Europe Using 14C Dating with Accelerator Mass Spectrometry

with Special Results for the Absolute Chronology of the Baden Culture," in *Cernavodă III–Boleráz: Ein vorgeschichtliches Phänomen zwischen dem Oberrhein und der unteren Donau* (Symposium Mangalia/Neptun, October 18–24, 1999), ed. P. Roman and S. Siamandi, Studia Danubiana, Series Symposia 2 (Bucharest: Vavila, 2001), 541–562; Tünde Horváth, S. Éva Svingor, and Mihály Molnár, "New Radiocarbon Dates for the Baden Culture," *Radiocarbon* 50, no. 3 (2008): 447–458.

4. Home on the Range

1. David W. Anthony, *The Horse, the Wheel, and Language: How Bronze-Age Riders from the Eurasian Steppes Shaped the Modern World* (Princeton, N.J.: Princeton University Press, 2007), 72.

2. Aeschylus, *Prometheus Bound*, trans. Robert Whitelaw (Oxford: Clarendon Press, 1907), lines 736–738.

3. Herodotus, *The Persian Wars*, trans. A. D. Godley, Loeb Classical Library 118 (Cambridge, Mass.: Harvard University Press, 1920), 4.46.3.

4. Ammianus Marcellinus, *Rerum gestarum* 31.2.20, http://perseus.uchicago .edu/perseus-cgi . . . %2031.2.20 (accessed October 9, 2014).

5. *The Mission of Friar William of Rubruck: His Journey to the Court of the Great Khan Möngke, 1253–1255*, trans. Peter Jackson (London: Hakluyt Society, 1990), 73–74.

6. "Romani Customs and Traditions: Death Rituals and Customs," *Patrin Web Journal: Romani Culture and History*, www.reocities.com/~patrin/death.htm (accessed October 9, 2014).

5. Wheels for Show

1. Stuart Piggott, *The Earliest Wheeled Transport: From the Atlantic Coast to the Caspian Sea* (Ithaca, N.Y.: Cornell University Press, 1983), 63.

2. László Tarr, *The History of the Carriage* (New York: Arco, 1969), 30.

3. E. B. Tylor, "Origin of the Plow and Wheel-Carriage," *Popular Science Monthly*, February 1881, 454 (emphasis added).

4. Ezra M. Stratton, *The World on Wheels; or, Carriages, with Their Historical Associations from the Earliest to the Present Time, Including a Selection from the American Centennial Exhibition* (New York: Published by the Author, 1878), 206.

5. M. A. Littauer and J. H. Crouwel, *Wheeled Vehicles and Ridden Animals in the Ancient Near East* (Leiden: Brill, 1979), 33.

6. Jared Diamond, *Guns, Germs, and Steel: The Fates of Human Societies* (New York: Norton, 1997), 171. I explore Diamond's general misunderstanding of domestication at length in *Hunters, Herders, and Hamburgers: The Past and Future of Human-Animal Relationships* (New York: Columbia University Press, 2004), chap. 5.

6. The Rise and Demise of the Charioteer

1. Since it is not known exactly where the bowl was found, its archaeological context cannot be dated. However, it is hard to imagine that any forger of antiques could have invented the scene.

2. M. A. Littauer and J. H. Crouwel, *Wheeled Vehicles and Ridden Animals in the Ancient Near East* (Leiden: Brill, 1979), 33.

3. Quoted in Robert Drews, *The End of the Bronze Age: Changes in Warfare and the Catastrophe ca. 1200 B.C.* (Princeton, N.J.: Princeton University Press, 1995), 125.

4. Ibid.

5. Plutarch, "Lucullus's Triumph over Mithridates, and His Luxurious Mode of Life," in *Readings in Ancient History: Illustrative Extracts from the Sources*, ed. William Stearns Davis, vol. 2, *Rome and the West* (Boston: Allyn and Bacon, 1913), 119.

6. Appian, "Pompey's Conquest of the East," in ibid., 123.

7. Flavius Vegetius Renatus, "Armed Chariots and Elephants," in *The Military Institutions of the Romans*, trans. John Clarke, www.digitalattic.org/home/war/vegetius/index.php#b321 (accessed October 10, 2014).

8. Homer, *Iliad*, trans. Richmond Lattimore (Chicago: University of Chicago Press, 1951), 23.364–372.

9. Quoted in Asko Parpola, "The Nāsatyas, the Chariot and Proto-Aryan Religion," 33, www.helsinki.fi/~aparpola/jis16–17.pdf (accessed October 10, 2014).

10. Einhard, "The Merovingian Family," in *The Life of Charlemagne*, trans. Samuel Epes Turner (New York: Harper, 1880), www.fordham.edu/halsall/basis/einhard.asp#The Merovingian Family (accessed October 11, 2014).

11. Heiko Steuer, "Archaeology and History: Proposals on the Social Structure of the Merovingian Kingdom," in *The Birth of Europe: Archaeology and Social Development in the First Millennium A.D.*, ed. Klaus Randsborg (Rome: "L'Erma" di Bretschneider, 1989), 109.

12. Quoted in Lynn White Jr., "The Origins of the Coach," *Proceedings of the American Philosophical Society* 114, no. 6 (1870): 417–428.

13. *Lancelot, the Knight of the Cart*, in *Chretien De Troyes: Arthurian Romances*, trans. W. W. Comfort (London: Everyman's Library, 1914), verses 247–398.

7. The Princess Ride

1. [John Kitto], "The Deaf Traveller—No. 5. Vehicles of Persia and Turkey," *Penny Magazine of the Society for the Diffusion of Useful Knowledge*, October 19, 1833, 407 (emphasis added).

2. *The Mission of Friar William of Rubruck: His Journey to the Court of the Great Khan Möngke, 1253–1255*, trans. Peter Jackson (London: Hakluyt Society, 1990), 73–74.

3. Ammianus Marcellinus, *Rerum gestarum* 31.2.20, http://perseus.uchicago.edu/perseus-cgi/citequery3.pl?dbname=LatinAugust2012&getid=1&query=Amm.%2031.2.20 (accessed October 9, 2014).

4. Ibid., 31.2.18.

5. 'Ata-Malik Juvaini, *The History of the World-Conqueror*, trans. John Andrew Boyle (Cambridge, Mass.: Harvard University Press, 1958), 1:212.

6. Thomas Moore, "Book II: Of the Travelling of the Utopians," in *Utopia*, trans. Ralph Robinson (New York: Collier, 1909–1914), oregonstate.edu/instruct/phl302/texts/morc/utopia-travelling.html (accessed October 11, 2014) (emphasis added).

7. William Goodman, *The Social History of Great Britain During the Reigns of the Stuarts: Beginning with the Seventeenth Century, Being the Period of Settling the United States* (New York: Colyer, 1843), 1:81.

8. The Carriage Revolution

1. Julian Munby, "From Carriage to Coach: What Happened?" in *The Art, Science, and Technology of Medieval Travel*, ed. Robert Bork and Andrea Kann (Farnham: Ashgate, 2008), 41–53.

2. "How Our Ancestors Traveled," *Temple Bar: A London Magazine for Town and Country Readers*, November 1871, 320, reprinted in *New York Times*, October 29, 1871.

3. Munby, "From Carriage to Coach," 43–53.

4. Quoted in László Tarr, *The History of the Carriage* (New York: Arco, 1969), 187.

5. Ibid., 211.
6. Ibid., 215. Note that Tarr does not specify what comfort he has in mind.
7. Ibid., 205.
8. Munby, "From Carriage to Coach," 50–51.
9. Technically, this pattern is called a logistic curve, and it is based on probability. "Innovators" are defined as individuals who are doing something highly improbable. In statistical terms, they fall three standard deviations below the mean adoption point of all people who eventually choose the innovation. "Early adopters" are doing something slightly less improbable and thus fall two standard deviations below the mean. Around two-thirds of all adopters fall into the bandwagon period. The "early majority" consists of those individuals who are one standard deviation below the mean, while the "late majority" falls one standard deviation above the mean. Finally, the "laggards" are improbably late and thus fall two to three standard deviations above the mean.
10. Thomas Asbridge, *The Greatest Knight: The Remarkable Life of William Marshal, the Power Behind Five English Thrones* (New York: HarperCollins, 2014), 48–49.
11. Flavius Vegetius Renatus, "How to Manage Raw and Undisciplined Troops," and "Armed Chariots and Elephants," both in *The Military Institutions of the Romans*, trans. John Clarke, http://www.digitalattic.org/home/war/vegetius/index.php#b308 and www.digitalattic.org/home/war/vegetius/index.php#b 321 (accessed October 10, 2014).
12. Quotes in Stuart Piggott, *Wagon, Chariot and Carriage: Symbol and Status in the History of Transport* (London: Thames & Hudson, 1992), 151, 153.

9. Four Wheels in China

1. *The Mission of Friar William of Rubruck: His Journey to the Court of the Great Khan Möngke, 1253–1255*, trans. Peter Jackson (London: Hakluyt Society, 1990), 55.
2. *Departure Herald* (Ming Dynasty [1368–1644]), National Palace Museum, Taiwan, www.npm.gov.tw/en/Article.aspx?sNo=04001151 (accessed October 13, 2014).

10. Rickshaw Cities

1. Greg Vore, "History and Development," in *Rickshaw Wallah*, www.rickshaw wallah.com/read-me/ (accessed October 13, 2014).
2. Saitō Toshihiko, *Jinrikisha* (Tokyo: Sangyō Gijutsu Senta, 1979); James Francis Warren, *Rickshaw Coolie: A People's History of Singapore, 1880–1940* (Singapore:

Singapore University Press, 2003); David Strand, *Rickshaw Beijing: City People and Politics in the 1920s* (Berkeley: University of California Press, 1993).

3. Owen Rutter, *Through Formosa: An Account of Japan's Island Colony* (London: Fisher Unwin, 1923), 214.

4. Katsushika Hokusai, *Fire Wheel* (ca. 1840), ukiyo-e.org/image/japancoll/p65 -hokusai-fire-wheel-8432 (accessed December 29, 2014).

5. Goble's claim is most thoroughly explored in F. Calvin Parker, *Jonathan Goble of Japan: Marine, Missionary, Maverick* (Lanham, Md.: University Press of America, 1990), chap. 17.

6. *Catalogue: Jinrikishas & Accessories* (*Jinrikisha katarogu*) (Tokyo: Daisuke Akiha, 1911), quoted in M. William Steele, "The Transfer of Rickshaw from Japan to East and Southeast Asia" (paper presented at the Fiftieth Congress of the Business History Society of Japan, Tokyo, September. 11–13, 2014), 2 (emphasis added).

7. Lauryn Noahr, "*Jinrikisha* in Meiji Japan," *Wittenberg History Journal* 37 (2008): 9.

8. Strand, *Rickshaw Beijing*, 20.

9. Since three men usually rotate pulling a rickshaw, the number of pullers may be close to eighteen thousand, according to Nancie Majowski, "Who Are the Rickshaw Wallahs?" June 7, 2008, and "Hand-Pulled Rickshaws and Kolkata's Image," March 17, 2008, GeoPedia, *National Geographic*, http://ngm.national geographic.com/geopedia/Kolkata_Rickshaws (accessed May 21, 2015).

10. Merritt Roe Smith and Leo Marx, eds., *Does Technology Drive History? The Dilemma of Technological Determinism* (Cambridge, Mass.: MIT Press, 1994).

11. Karl Marx, *The Poverty of Philosophy* (Moscow: Progress Publishers, 1955), 49.

12. Rosalind Williams, "The Political and Feminist Dimensions of Technological Determinism," in *Does Technology Drive History?* ed. Smith and Marx, 225.

11. The Third Wheel

1. David A. Fisher Jr., "Improvement in Furniture-Casters," US Patent 174794 A, filed March 1, 1876, and issued March 14, 1876, www.google.com/patents/ US174794 (accessed October 15, 2014) (emphasis added).

Glossary

ACKERMANN STEERING Invented in Munich in 1817 by Georg Lankensperger and patented in England by his friend Rudolf Ackermann, Ackermann steering forms the basis for all modern steering designs for motor vehicles. Each front wheel pivots separately at the end of an axle that does not itself pivot. Rods behind the axle linking the two wheels ensure that each of them turns at a somewhat different angle according to whether the wheel is on the inside or the outside of the turning radius.

BALLAST Small stones that form a stable and well-drained foundation for railroad tracks. *See also* MACADAM.

BATH CHAIR Invented by James Heath in the English resort city of Bath in 1750, the Bath chair was pulled or pushed by an attendant and resembled a modern wheelchair. The attendant or passenger controlled the steering of a small, single front wheel—not a caster—by means of a TILLER.

BATTEN A short piece of wood used to reinforce the connection between the parts of a solid wooden wheel made of several pieces.

BREAST STRAP A way of harnessing a horse by means of traces or shafts that attach to a strap that goes around the animal's breast.

BUGGY The general nineteenth-century American word for a horse-drawn passenger vehicle. Synonym of CARRIAGE and COACH. The term was also used broadly, as in "baby buggy."

CARRIAGE The standard English word for a horse-drawn passenger vehicle, later extended to railroad cars. Synonym of COACH.

CARRIAGE REVOLUTION The transition in European transportation between 1450 and 1650 from four-wheeled passenger vehicles being used almost exclusively by aristocratic women to their becoming the standard means of transport of the upper class.

CART A vehicle with two wheels placed side by side. The cart was the most common sort of wheeled vehicle worldwide from 2000 B.C.E. until 1900 C.E.

CASTER A wheel that both turns on an axle and pivots in a socket located above, and offset either before or behind, the axis of rotation of the axle. Apparently invented in the eighteenth century, the caster afforded a new means of steering because lateral pressure caused it to turn to accommodate the pressure.

CASTER ANGLE The degree of offset between a caster's axis of rotation and its axis of pivoting.

COACH A word of Hungarian origin (*kocsi*), but adopted into many other languages, that became the general term for stylish four-wheeled passenger vehicles in central and western Europe during the fifteenth and sixteenth centuries. Synonym of CARRIAGE.

CROSSTIES Pieces of wood or concrete placed at close intervals underneath and perpendicular to railroad tracks to keep them aligned and at a constant gauge, or distance apart, from each other.

ELLIPTICAL LEAF SPRING A group of several arced strips of spring metal clamped together. One group can be used as a vehicle spring, but more commonly a downward-facing arc is placed on top of an upward-facing arc to form a CARRIAGE spring that is roughly elliptical in shape.

FACTORY TRUCK (OR FACTORY CART) A flat platform on wheels used to shift loads in factories in the late nineteenth and early twentieth centuries. The factory truck usually had two sizable wheels (or one

WHEELSET) in the middle, with one or two smaller casters before and behind to facilitate steering.

FELLOE (OR FELLY) In a spoked wheel, one of several curved pieces of wood that, pieced together, form the RIM of a wheel and into which the outer end of each SPOKE is inserted.

FLANGE The lip on the outer (or sometimes inner) edge of a railroad wheel that goes over the edge of the track and thereby prevents the wheel from derailing.

FOUR-IN-HAND DRIVING The difficult skill of controlling four horses while driving a CARRIAGE.

FULL LOCK In a WAGON or CARRIAGE with a PIVOTING FRONT AXLE, the limit on the degree of turn that can be accomplished as a result of contact between the RIM of the wheel and the body of the vehicle.

HANDCART A CART pulled by a human instead of a draft animal.

HANSOM CAB A type of two-wheeled CARRIAGE pulled by a horse, with the driver sitting in an elevated position behind the covered passenger compartment. The hansom cab was commonly used as a vehicle for hire in London and elsewhere in the nineteenth century.

HOBNAIL In ancient times, a large bronze or iron nail with a rounded head that was pounded in a series around the RIM of a wooden wheel to protect it from wear and provide traction.

HORSE COLLAR A type of CART or WAGON harness in which a padded collar circles a horse's neck and the shafts or traces that connect the horse to the vehicle attach to hooks, called hames, midway down the sides of the collar. Apparently invented in China and adopted in Europe in the post-Roman period, the horse collar transferred the pulling load to a horse's shoulders from its throat, where the YOKE harness of ancient times placed it.

HUB In a spoked wheel, the central piece into which the inner end of each SPOKE is inserted. In a wooden wheel, the hub also serves as the NAVE.

MACADAM Road-building technique devised by John McAdam in the early nineteenth century. McAdam discovered that a road surface of small stones with sharp edges compacted as traffic moved over it, while one made of rounded stones dispersed. Later engineers favored spreading a binder, either tar or stone dust mixed with water, over the stones. Thus the words "macadam" and its derivative "tarmac" (for tar-macadam) came to be applied to a variety road surfaces.

NAVE In a wooden wheel, the central portion through which the axle is inserted. The nave, which may be made from a different piece of wood from the rest of the wheel, usually is much thicker than the rest of the wheel because a sleeve at least six inches thick prevents the wheel from wobbling on the axle. In a spoked wheel, the nave receives the inner ends of the SPOKES and thus becomes a HUB.

ORDINARY *See* PENNY-FARTHING.

PALANQUIN A small chamber or litter, usually intended for a single passenger, carried by two or more bearers or suspended between two animals. In Japan, the passenger rode beneath the poles on the bearers' shoulders. In China, the chamber extended above and below the poles. In the Middle East, it rested on the poles. Synonym of SEDAN CHAIR.

PENNY-FARTHING (OR ORDINARY) A type of bicycle that gained great popularity in the late nineteenth century. Because the penny-farthing does not have gears, its front wheel is very large so that it can cover a substantial distance as it rotates once with every circuit of the pedals. By contrast, the back wheel, which helps balance the rider, is small, reminding onlookers of the difference between a large English penny and a tiny English farthing.

PILLION The cushioned seat behind a horse saddle on which a second rider can sit.

PIVOTING FRONT AXLE To facilitate turning, four-wheeled vehicles before the invention of ACKERMANN STEERING had a front axle that pivoted underneath the vehicle's body. The pivot might be a single point or be accomplished by a circular device.

POSTILION The rider of one horse in a team of horses drawing a CARRIAGE or WAGON. The postilion helped the driver control the horses.

PUSH RAILROAD In Japan, and in Taiwan when it was ruled by Japan, a freight or passenger railroad whose operating power came from one or two men pushing the small cars.

RAILWAY TIME The synchronization of clocks instituted in the mid-nineteenth century by the Great Western Railway in order to prevent accidents, especially when trains ran on a single track. Railway time paved the way for the formal adoption of standard time within geographically delimited time zones and thus became a byword for precise scheduling.

RICKSHAW A two-wheeled, human-pulled passenger vehicle invented in Japan in 1869, whose name derives from the Japanese word *jinrikisha* (human-powered vehicle). The rickshaw commonly had a folding top and an ELLIPTICAL LEAF SPRING SUSPENSION. The use of rickshaws spread rapidly throughout East and South Asia and even to East Africa.

RIM The strip of protective material, usually metal, that is secured to the outer edge of a wooden wheel. A rubber rim is called a tire.

ROLLER A thick pole that is placed under a very heavy load, but not attached to the load. The load rests on several rollers as it moves forward, and new poles are added in front as others become free behind.

ROLLING RESISTANCE A determinant of efficiency in vehicle design. The amount of force needed to set a vehicle in motion and keep it moving is partly determined by the contact between the wheels and the traveling surface. Different combinations—iron wheels on iron rails with little resistance, as opposed to inflated rubber tires on concrete with greater resistance, for example—produce different degrees of rolling resistance.

SEDAN CHAIR *See* PALANQUIN.

SLEDGE A conveyance that moves on runners, like a sled, but is not dependent on snow. Not requiring an artificially leveled surface, the sledge was used in some parts of the world before the invention of the wheel and was still commonly found on farms in the early twentieth century.

SPOKE The stick, rod, or wire that goes between the HUB of a wheel and its outer edge. Originally, the spoke was a separate wooden piece that was inserted into the hub at one end and into a FELLOE at the other.

SUSPENSION Any of several ways of improving the comfort of passengers by allowing a vehicle to sway in response to the unevenness of a road or, by using springs, to absorb jolts and bounces.

TILLER Like the steering handle attached to the rudder of a sailing vessel, the tiller was a handle attached to the front wheel or wheels of some early automobiles or to a hand-pulled vehicle like the BATH CHAIR.

TRAMWAY Parallel rails initially used inside, and eventually outside, mines to provide a guide-path for ore-cars or railcars. Early tramways were propelled by horses or humans, but the word "tram" or "streetcar" came to be applied to electric light-rail vehicles.

TRAVOIS A simple device that can be attached to dogs and humans that was used for moving loads in pre-Columbian America and was adapted for use by a horse after European contact. Two sticks are fastened together at one end and tied to the puller. At the far end, they spread widely and are in contact with the ground. The load is placed over the *V*-shaped space between the sticks on a blanket or another suitable connector.

WAGON A four-wheeled vehicle. After the CARRIAGE REVOLUTION, which ushered in improvements in SUSPENSION and braking, the wagon often replaced the CART for hauling freight, crops, and farm implements.

WHEELSET The earliest type of wheel. The wheelset consists of two wheels fixed to the ends of an axle. The axle and the wheels rotate as a unit underneath the vehicle, where the assembly is held in place by brackets with holes in them or by grooves or pegs.

WHEELWRIGHT An artisan who is skilled in manufacturing wheels.

YOKE A device for harnessing a pair of draft oxen that consists of a crossbar at the end of a beam extending from the front of a plow, CART, or WAGON. Either end of the yoke rests on the neck of one of the oxen and is kept in place by pegs sticking down from it. The pulling load is borne by the hump-like protrusion of the thoracic vertebrae that is characteristic of bovine animals. When adapted in ancient times for horses and other equids, which do not have protruding vertebrae, the yoke was held in place by a strap around the neck, and the pulling pressure was borne by the animal's throat. The BREAST STRAP and the HORSE COLLAR made the yoke obsolete for horses and came to be called modern harnessing.

Further Reading

This book has emphasized new ideas and a broad historical perspective. It has steered clear of many scholarly debates on more localized or technical issues, but readers who want to delve further into the history of wheeled transport may wish to consult the following works.

General

Lay, M. G. *Ways of the World: A History of the World's Roads and of the Vehicles That Used Them*. New Brunswick, N.J.: Rutgers University Press, 1992.

Piggott, Stuart. *Wagon, Chariot and Carriage: Symbol and Status in the History of Transport*. London: Thames & Hudson, 1992.

Treue, Wilhelm, ed. *Achse, Rad und Wagen: Fünftausend Jahre Kultur- und Technikgeschichte*. Göttingen: Vandenhoeck & Ruprecht, 1986.

Ancient

Anthony, David W. *The Horse, the Wheel, and Language: How Bronze-Age Riders from the Eurasian Steppes Shaped the Modern World*. Princeton, N.J.: Princeton University Press, 2007.

Anthony, David W., with Jennifer Y. Chi, eds. *The Lost World of Old Europe: The Danube Valley, 5000–3500 BC.* Princeton, N.J.: Princeton University Press, 2010.

Drews, Robert. *The End of the Bronze Age: Changes in Warfare and the Catastrophe ca. 1200 B.C.* Princeton, N.J.: Princeton University Press, 1995.

Fansa, Mamoun, and Stefan Burmeister, eds. *Rad und Wagen: Der Ursprung einer Innovation. Wagen im Vorderen Orient und Europa.* Mainz: Philipp von Zabern, 2004.

Littauer, M. A., and J. H. Crouwel. *Wheeled Vehicles and Ridden Animals in the Ancient Near East.* Leiden: Brill, 1979.

Mallory, J. P., and D. Q. Adams. *The Oxford Introduction to Proto-Indo-European and the Proto-Indo-European World.* Oxford: Oxford University Press, 2006.

Piggott, Stuart. *The Earliest Wheeled Transport: From the Atlantic Coast to the Caspian Sea.* Ithaca, N.Y.: Cornell University Press, 1983.

Domestic Animals Used for Draft

Bulliet, Richard W. *The Camel and the Wheel.* Cambridge, Mass.: Harvard University Press, 1975.

——. *Hunters, Herders, and Hamburgers: The Past and Future of Human-Animal Relationships.* New York: Columbia University Press, 2004.

Langdon, John. *Horses, Oxen and Technological Innovation: The Use of Draught Animals in English Farming from 1066–1500.* Cambridge: Cambridge University Press, 1986.

Carriages

Felton, William. *A Treatise on Carriages; Comprehending Coaches, Chariots, Phaetons, Curricles, Gigs, Whiskies, &c Together with their Proper Harness. In which the Fair Prices of Every Article are Accurately Stated.* London: Printed for and Sold by the Author, 1794. https://archive.org/search.php?query=%22william%20felton%22.

"How Our Ancestors Traveled." *Temple Bar: A London Magazine for Town and Country Readers,* November 1871, 320–338.

Kugler, Georg J. "Die Kutsch vom Beginn des 18. Jahrhunderts bis zum Auftreten des Automobils." In *Achse, Rad und Wagen: Fünftausend Jahre Kultur- und Technikgeschichte,* ed. Wilhelm Treue. Göttingen: Vandenhoeck & Ruprecht, 1986.

Tarr, László. *The History of the Carriage.* New York: Arco, 1969.

Thrupp, G. A. *The History of Coaches.* London: Kerby & Endean, [1899]. ia600406. us.archive.org/33/items/historyofcoachesoothru/historyofcoachesoothru.pdf.

Tristram, W. Outram. *Coaching Days and Coaching Ways.* London: Macmillan, 1888. http://www.archive.org/stream/coachingdaysandootris#page/n9/mode/2up.

Trains and Automobiles

Caro, Robert A. *The Power Broker: Robert Moses and the Fall of New York*. New York: Knopf, 1974.

Kemper, Steve. *Reinventing the Wheel: A Story of Genius, Innovation, and Grand Ambition*. New York: HarperBusiness, 2005.

McGowan, Christopher. *Rail, Steam, and Speed: The "Rocket" and the Birth of Steam Locomotion*. New York: Columbia University Press, 2004.

Schivelbusch, Wolfgang. *The Railway Journey: The Industrialization of Time and Space in the Nineteenth Century*. Berkeley: University of California Press, 1986.

Warner, Sam Bass, Jr. *Streetcar Suburbs: The Process of Growth in Boston, 1870–1900*. 2nd ed. Cambridge, Mass.: Harvard University Press, 1978.

Zeller, Thomas. *Driving Germany: The Landscape of the German Autobahn, 1930–1970*. New York: Berghahn, 2006.

Central Asia

Hildinger, Erik. *Warriors of the Steppe: A Military History of Central Asia, 500 B.C. to 1700 A.D.* New York: Sarpedon, 1997.

Mallory, J. P., and Victor H. Mair. *The Tarim Mummies: Ancient China and the Mystery of the Earliest Peoples from the West*. London: Thames & Hudson, 2000.

Rolle, Renate. *The World of the Scythians*. Translated by F. G. Walls. Berkeley: University of California Press, 1980.

East Asia

Saitō Tosihiko. *Jinrikisha* [in Japanese]. Tokyo: Sangyo Gijutsu Senta, 1979.

Strand, David. *Rickshaw Beijing: City People and Politics in the 1920s*. Berkeley: University of California Press, 1993.

Warren, James Francis. *Rickshaw Coolie: A People's History of Singapore, 1880–1940*. Singapore: Singapore University Press, 2003.

Index

Numbers in italics refer to pages on which illustrations appear.

archaeological evidence (*continued*)
Indus Valley clay cart model as, *74*; from Mesopotamia, 93, *95*, 95–96, 99–100, 103–106, *104* (*see also* Mesopotamia; Sumer); for Scythian mobile homes, 87, *88*; for spoked-wheel development, 82, 82–85, *83*, *85*; wheeled toys as, 37–38, *38*, 53–54, *54*; women not seen in, 133. *See also* Copper Age; Egypt, ancient

Asia. *See* Central Asia; China; India; Japan; Southeast Asia; Taiwan

asses (mules), 122–124

Assyrian bas relief, *40*

automobiles: advertisements for, 31, *34*; Benz models of, 10, *11*, 12, 13, *14*; cost efficiency of, 24; draft animals' fate sealed by, 185, 219; electric starter in, 30–32; in film and television, 35–36; footprint of, 197; and image of the driver, 31; independently rotating wheels on, 3; in Japan, 189–190; as personal/economic statements, 4; rise of, 219; rolling resistance of, 20–21, 24; speedometers for, 29; and trains and light rail, 19–20, 24, 27; U.S. statistics on, 19. *See also* highways; road design and construction; roads; trucks

axles: early wooden, *64*; and fixed wheels, 1, 234 (*see also* fixed-wheel vehicles; wheelsets); and independently rotating wheels, 1, 72–73 (*see also* wheels: independently rotating); pivoting front, 12, 75, 129, *130*, 148, 167–169, 173–174, 231, 232; on single-wheelset oxcarts, 70; and steering/turning, 2–3, 12–13. *See also* hub; nave

Babbage, Charles, 29

baby carriages and strollers, 2, 185, 187, 192, 206, 211–212, 217

Baden Culture. *See* Boleráz clay mugs

Bath chairs, 8, *9*, 192, 213, 218, 229, 234

battens, 84, 106, *107*, 229

Beatrice of Naples, 149–150

Beaumont, Huntingdon, 16

Belušić, Josip, 29

Benjamin, Samuel G. W., 5

Benz, Karl, 10, 12, 13

Benz Patent Motorwagen (1885), 10, *11*, 12

Benz Velo (1894), 13, *14*

Bhagavad Gita, 119

bicycles: cycle-rickshaw, 187, 200, *201*, *202*, 209; evolution and popularity of, 185, 205, 208–210; "ordinary" ("penny-farthing"), 208, *209*, *210*, 232; steering of, 208, *210*, 211; and technological invention, 208, 219; use of, for racing, 30, 208–209, *210*

bit (horse tack), 111

Black Sea plain: and evolution of the chariot, 115–117, 121–122; nomadism on, 79, 87, 110, 113, 133, 137–138, 161; and PIE, 79–81 (*see also* Proto-Indo-European); prehistoric climatic and geologic changes on, 75–77, 84, 138; society on, 99; wagons and wagon burials on, 71–79, *72*, 102–103, 113, 145–146; wild horses on, 108

Boleráz clay mugs (wagon models), 58–69, *59*, *60*, *65*, 80

Bondár, Mária, 58–59, 62–63, 65, 67. *See also* Boleráz clay mugs

bows and arrows, 117–119, *118*, 121

brakes and braking, 16–17, 35, 166

Broady, Steve, 35

Bronocice pot, 62, *63*
brouette (French passenger cart), 192, *193*, 200
buggies, 230. *See also* carriages; coaches

caltrops, 120
Camel and the Wheel, The (Bulliet), 43
camels, 42–43, *110*, 110–111, 125–126, 134–135, *135*
cannon, 156–157, *158*, 159, 171, 175. *See also* gunpowder, guns, and cannon
carbon-14 and carbon dating, 52–53, 61
Carpathian wheel-origin theory, 51, 53–58, 99, 113; Boleráz clay mugs as evidence for, 58–69, *59*, *60*, *65*
carriage revolution, 4–5, 166; adoption curve of, 154–155, *155*; causes of, 148–149; and changing military tactics, 159–161; dates of, 150, 230; definition of, 230; and gender, 132; Hungary as origin point of, 152–154, *153*, 161; and image of the driver, 30, 147, 151–152, 160–161; and origin and use of "coach" as term, 4, 152–154, 162, 230; and road conditions, 150–151; and transition to steam and gasoline power, 162; and urban street design, 26; westward spread of, 152, 161–162. *See also* carriages; coaches
carriages: accidents involving, *151*; bridal, *149*, 149–150; definition of, 230; four-in-hand driving of, 30–31, *31*, 231; hansom cabs as, 196, 197, *197*; in Japan, 188, *188*; papal, *150*; as personal/economic statement, 4, 30, 147, 161; rickshaws as low-tech response to, 204 (*see also* rickshaws); size of wheels on, 216 (*see also* wheels:

size of); suspension of, 148, *149*, 153, 192–194, *193* (*see also* suspension systems); as symbol of cultural/technological advancement, 4–5; three-wheeled steam, 7–10, *8*, *9*, *10*; two-wheeled, *32*; types of, 4; women as drivers of, 31, *32*. *See also* carriage revolution; coaches
cars. *See* automobiles
carts: *brouettes* (French passenger carts), 192, *193*, 200; in China, 177; definition of, 230; economic efficiency of, 42–43; first appearance of, 85; and four-wheeled vehicles, 3–4, 74–75, 79, 86, 102, 145, 161–163 (*see also* steering); harnesses for (*see* harnesses); human-drawn, 39–40, *40*, 231 (*see also* mining: and mine-cars; rickshaws); independently rotating or fixed wheels on, 6–7; Indus Valley clay model of, *74*; in Japan, 44–49, *47*, 187 (*see also* Japan); in Middle East, 42–43; not adopted by some societies/civilizations, 37–43, 165; pushcarts and handcarts, 206, *206*; Roman load limits for, 22; scorned as vehicles for men, 128–132, 142, 144, 145; size of wheels on, 216 (*see also* wheels: size of); in Sumer, 103, *104*, 106–108, *107*, 111–112, 116; use of, in factories, 213–214, *215*, 217, 230–231 (*see also* dollies); use of, for laundry and refuse, 66, *66*, 211; use of, on medieval European farms, *146*; use of, by nomads, 87 (*see also* nomadism, wheeled); use of, for shopping, 3, 185, 211–212; value of, 39, 41. *See also* chariots; oxcarts; wagons

palanquins, 134–135, *135*, 144, 172, *173*, *174*, 194, *195*, 232. *See also* sedan chairs

Pallas, Peter Simon, 89–90, *90*

passenger vehicles. *See* automobiles; carriage revolution; carriages; carts; railroads; rickshaws; wagons

Patrick, Danica, 33

pedestrians, 25–27

"penny-farthing" ("ordinary") bicycles, 208, *209*, *210*, 232. *See also* bicycles

Perry, Matthew C., 190–191

Persia, 5, 43, 134–135, *135*

Persian Empire, 124–125, *126*

PIE. *See* Proto-Indo-European

Piggott, Stuart, 51–53, 61, 93, 98, 99, 105

pillion, 143, 232

pivoting front axles: and carriage revolution, 148; and Chinese vehicles, 173–174; and full lock, 12, 129, 231; function of, 12, 75, 232; and Mongol wagons, 167–169; on Roman passenger wagons, 129, *130*. *See also* axles

Pompey (Roman general), 120

porte cochere, 147

postilion, 30, *142*, *143*, 232

pre-Columbian peoples, 87, 165, 234; toy dog of, 37–39, *38*, 62, 218

processions, wheels used for, 99–105, *101*, *102*, 176, *177*, *178*, 179. See also *Departure Herald*

Proto-Indo-European (PIE; language), 79–82, 137

pushing or pulling, 186–187, 218–219

push railroad, 189, *190*, 232

racing: of automobiles, 30, 33; of bicycles, 30, 208–209, *210*; of carts,

111–112; of chariots, 121–123, 126–127; of locomotives, 29

railroads: cost efficiency of, 24; as culmination of carriage revolution, 162; and evolution of steam locomotive, 7–10, 13–15; folk heroes of, 35; hand-pushed, 189, *190*, 232; in India and Africa, 24–25; in Japan, 188, 189, *190*, 232; as mass transit, 20; pivoting-wheel assembly of, 15; rights-of-way and maintenance of, 18, 26–27; and roads and road vehicles, 19–20, 24, 35–36; rolling resistance of, 20, 24; scheduling (timetables) of, 28–29, 233; speed/velocity of, 29, 35; turning of, 2–3, 18; wheelsets on, 2, *2*, 15. *See also* rail tracks; *specific railroad companies*

rail tracks: crossties and ballast on, 22, *23*, 229, 230; evolution of, 15–18, *17*; grade of, 16–18; for mine-cars, 6–7, 15–18, 234 (*see also* mining: and mine-cars); for push-car lines, 189; and rights-of-way, 18, 26–27; and urban planning, 26–27; and wheel flanges, 16, 231. *See also* railroads

railway time, 28, 233

Ramses II, 117–118, *118*

refuse cart, 66, *66*

religion: and chariots, *115*, 115–116, 121–124; and horses, 125; of Hussites, and wars against Catholics, 147, 157–160, *158*, 162, 175; and wheels, 99–105, *101*, *102*, 117

Rickshaw Beijing: City People and Politics in the 1920s (Strand), 187

Rickshaw Coolie: A People's History of Singapore, 1880–1940 (Warren), 187

wagons (*continued*)

three-wheeled steam, 7–8, *8*; value of, 39, 41; Viking, *82*, 83; wheel size of, 216 (*see also* wheels: size of). *See also* carts; horses; oxen; wagon burials

Wanli Emperor, 180, 182

war: bow and arrow as charioteer's weapon in, 117–120, *118*; in central Europe, and technological advances, 147; chariots superseded by cavalry in, 120–121, 124–125, 156; chariot warriors glorified in, 117–119, *118*, 122, 137; Egyptian chariots in, 41; and evolution of war chariot, 111–112, *115*, 115–121; gunpowder, guns, and cannon in, 156–157, *158*, 159–160, 162, 171; Hussites' tactics in, 157–160, *158*, 162, 175; infantry in, 105, 119–120, 157, 171; and noblemen's attitudes toward horses, carriages, 155–156 (*see also* knights); and origin of words "howitzer" and "pistol," 160, 162; Roman chariots in, 102; Sumerian wagons and carts in, 103–105, *104*

Warren, James Francis, 187

Watt, James, 13

wheelbarrows: and casters, 219; in China, 43–44, *45*, 114, 172, *173, 174*, 177, 204; in Europe, 44, 114

wheelchairs: casters on, 2, 205, 206, 211–212, 213, *214*; and social attitudes toward pushing/pulling, 187; steering/turning of, 3

wheels: from Bronze Age, 84–85, *85* (*see also* chariots; wheels: spoked; *specific cultures*); design and construction of

(*see* wheels: independently rotating; wheels: solid; wheels: spoked); independently pivoting, 3, 12–13, *14*, 229 (*see also* Ackermann steering); types of, 1–3 (*see also* casters; wheels: independently rotating; wheelsets)

INDEPENDENTLY ROTATING: archaeological evidence for, 71–79, *72*; construction of, 70, 72–73; dangers of, 6, 15; definition of, 1; durability of, 7; invention of, 1–2, 6, 71, 80; nave of, 72, 73, *74*; and PIE, 80; road vehicles' use of, 2; in Sumer, 106–108, *107*; and wheelsets, 3, 149; widespread adoption of, 71–75, *72*, 113 (*see also* Black Sea plain). *See also* wheels: independently pivoting

INVENTION OF: on Black Sea plain (*see* Black Sea plain); in Carpathian mines, 51, 53–58, 105–106, 113; challenges of determining, 51–53; earliest wheels as archaeological evidence of, 6, 52–53; and independent technological development, 105–106, 108, 219; in Mesopotamia, 51–53, 61, 93, *95*, 95–98, *97*, 105–106; roller-origin theory of, 96–98, *97*; spread as idea rather than technology, 105–106

SIZE OF: on Boleráz clay mugs, 62–66; on carts, wagons, and carriages, 216; of casters, 216–217; on Chinese wagon, 170; earliest wheels, 6, *64*, 64–67, 73; and pivoting axles, 167–169; spoked wheels, 85; and thickness and weight, 72–73; and tree diameter, 84